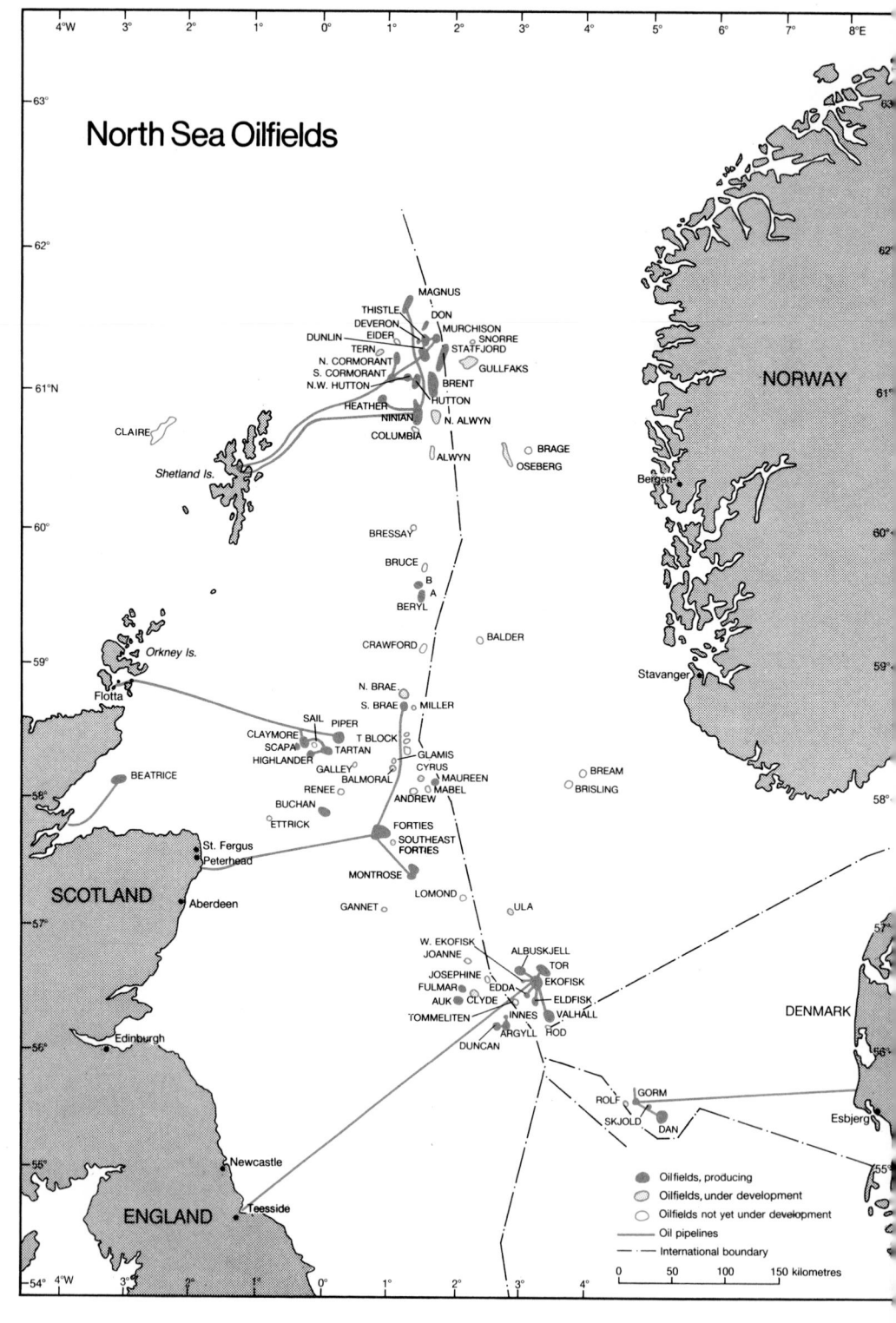

North Sea Oilfields

MAGNUS

THISTLE
DEVERON
DON
EIDER
MURCHISON
DUNLIN
SNORRE
TERN
STATFJORD
N. CORMORANT
GULLFAKS
S. CORMORANT
N.W. HUTTON
BRENT
HEATHER
HUTTON
NINIAN
N. ALWYN
COLUMBIA

CLAIRE

ALWYN

Shetland Is.

BRAGE
OSEBERG

Bergen

NORWAY

BRESSAY

BRUCE

B
A
BERYL

Orkney Is.

Flotta

CRAWFORD

BALDER

Stavanger

N. BRAE
S. BRAE
MILLER

SAIL
PIPER
CLAYMORE
T BLOCK
SCAPA
TARTAN
HIGHLANDER
GALLEY
BALMORAL
GLAMIS
CYRUS
MAUREEN
RENEE
MABEL
ANDREW
BREAM
BRISLING

BEATRICE

BUCHAN
ETTRICK

FORTIES
SOUTHEAST
FORTIES

St. Fergus
Peterhead

MONTROSE

SCOTLAND

Aberdeen

GANNET

LOMOND

ULA

W. EKOFISK
JOANNE
ALBUSKJELL
TOR
JOSEPHINE
FULMAR
EKOFISK
AUK
EDDA
CLYDE
ELDFISK
TOMMELITEN
INNES
VALHALL
ARGYLL
HOD
DUNCAN

DENMARK

Edinburgh

ROLF
GORM
SKJOLD
DAN

Esbjerg

Newcastle

ENGLAND

Teesside

Oilfields, producing
Oilfields, under development
Oilfields not yet under development
Oil pipelines
International boundary

0   50   100   150 kilometres

The Market for North Sea Crude Oil

The Market for North Sea Crude Oil

ROBERT MABRO
ROBERT BACON
MARGARET CHADWICK
MARK HALLIWELL
DAVID LONG

Published by the Oxford University Press
for the Oxford Institute for Energy Studies
1986

Oxford University Press, Walton Street, Oxford OX2 6DP

Oxford   New York   Toronto
Delhi   Bombay   Calcutta   Madras   Karachi
Petaling Jaya   Singapore   Hong Kong   Tokyo
Nairobi   Dar es Salaam   Cape Town
Melbourne   Auckland

and associated companies in
Beirut   Berlin   Ibadan   Nicosia

Oxford is a trade mark of Oxford University Press

British Library Cataloguing in Publication Data

The Market for North Sea Crude Oil
1. Offshore gas industry—North Sea
2. Offshore oil industry—North Sea
I. Mabro, Robert
333.8'23'0916336   HD9581.N642
ISBN 0-19-730001-4

Map: Petroleum Economist
Typeset by Burgess & Son (Abingdon) Ltd.,
in 11 on 12 point Baskerville
and printed in Great Britain by
Biddles Limited, Guildford

PREFACE

The North Sea has become an important source of world oil supplies and the market for North Sea crudes is now playing an active role in international petroleum trade. In April 1985, the North Sea, despite its smaller reserves and its higher development costs, produced more oil than Saudi Arabia; a temporary anomaly which raises however interesting questions about differences in policy and behaviour between aggressive newcomers and old-established leaders. Since 1982 or 1983, the market for North Sea crudes (often referred to as the Brent market) has been playing a 'barometric' role and has been influencing the process of price determination in the OPEC region.

Students of the world petroleum market cannot afford to ignore the North Sea. Its significance relates to at least three major factors.

First, the North Sea has been growing rapidly during a period in which world demand for oil has been stagnant. It has displaced OPEC supplies and, together with other emerging producers, presented a challenge to OPEC's role as the administrator of crude oil prices in international trade.

Secondly, the North Sea is located in the centre of a major refining and consuming area, namely North West Europe. This locational factor has transformed certain structural features of the oil industry. It has consolidated the Atlantic basin into a single oil trading area with close ties between the eastern shores of America and the western parts of Africa and Europe, and contributed to the weakening of the traditional links between the European market and the Gulf producers.

Thirdly, the development of the North Sea has led to the emergence of highly active spot and forward markets for crude oil. Competitive forms of trading, which had expanded in the USA after deregulation, penetrated the Eastern Hemisphere through the North Sea. There has been a multiplication in the number of oil buyers and sellers and a diversification of the type of companies engaged in petroleum trade. These new agents and these new modes of trading tend to introduce greater competition to other parts of the world petroleum market. These increasingly pervasive phenomena raise fundamental questions about future oil price behaviour and more specifically the troublesome question of price stability.

Since its foundation in 1983, the Oxford Institute for Energy Studies has been engaged in advanced economic research on the structure and operations of the world petroleum market. As this work progressed the research staff became increasingly aware of the need to investigate the significance of the North Sea and the market for North Sea crudes in some depth.

We also felt that many companies and governments, oilmen and expert observers of the industry were interested, albeit in different ways and from a variety of points of views, in the same or similar sets of questions. This perception, which crystallized in the minds of three of us at the Institute one day in late August 1984, led us to formulate a research project and to seek financial support from corporate members of the Institute and from other institutions.

This is the place to express our gratitude to the sponsors which we have pleasure listing in alphabetical order:

Amerada Hess
APICORP
British Petroleum
Britoil
Department of Energy (UK)
Elf-Aquitaine
Esso Europe
EEC
Mobil
OPEC Secretariat
Pemex
Petromin
Qatar General Petroleum Corporation
Royal Ministry of Petroleum and Energy (Norway)
Shell Oil (US)
Shell UK
Statoil
Sun Oil
Svenska Petroleum
Veba Oel

We would also like to acknowledge a grant from the ESRC for a research programme on the world petroleum market. This research provided us with the broader framework to which the present book relates as an in-depth case study.

The study was conducted by an Institute team led by Robert Mabro. The members of the team include academic economists and professionals with experience of the oil industry. Robert Bacon, University Lecturer in Econometrics at Oxford, did most of the econometric work in co-operation with Mark Halliwell. David Long, now on secondment to the Institute from BP, and Margaret Chadwick covered most of the oil industry aspects. Peter Beck, formerly head of Corporate Planning, Shell UK, applied incisive critical skills to every working paper and every draft of the original report. Brian Haskell-Thomas interviewed a large number of oil executives and oil traders in the UK and Penny Walker, on sabbatical from General Electric, performed the same task in New York, Washington and Houston. Professor Alexander Kemp advised us on fiscal matters and provided us with estimates of taxes per field for the UK. Ali Khadr contributed on certain points of economic theory. Jayne Dexter assisted Long in the thankless task of processing numbers and surveying every issue of *Lloyd's List* and the *Shetland Times* for data on shipments. Nancy Gillespie, Yanna Zour and Clare Davison were occasionally employed as research assistants. Margaret Ko typed most of the final draft. Ann Davison organized the editing of the original report; Karen Exley arranged the production of the typescript and the various sponsor meetings, and Christine Hooper helped the three of them in various tasks. David Guthrie sub-edited the typescript for the purposes of publication and directed the production of the book.

This book was put together by Robert Mabro using working papers written by members of the team. The five other members of the core team, Bacon, Beck, Chadwick, Halliwell and Long, read every word of the final draft, and whenever any one of them dissented the issue was discussed and the text amended. Having worked together for the best part of nine months in the noisy intimacy of an open plan office the main researchers can no longer identify the origin and authorship of the ideas and conclusions of the study. They are thus happy to accept collective responsibility, although the ultimate blame for errors and for views that are clearly qualified as an individual opinion must rest with the director of the project.

The usual disclaimers apply. The Oxford Institute for Energy Studies is an autonomous and responsible research body and the results and conclusions of this study do not involve in any sense the corporate Members of the Institute, their governments or any affiliated organizations. *The Market for North Sea Crude Oil* is an official Institute study; any other work on the North Sea carried out at the Institute by research or visiting fellows in their individual capacity falls outside the scope of this study. The Institute encourages independent research and respects the full academic freedom in publication and debate of its staff, but is committed only to work published under its own imprint.

This study would not have been possible without the help of persons and institutions who kindly agreed to answer our questions and to provide us with statistical material and other information. We are particularly grateful to *Petroleum Argus* for giving access to their invaluable quarry of primary data, to Shell and Esso for a compilation of oil price statistics, to BP for access to *Platt's Crude Oil Marketwire*, and to Rosemary McFadden for supplying all the numbers and the information about the New York Mercantile Exchange. We interviewed seventy-three persons in the UK, the USA, Norway, Belgium and France. However we shall not list them here lest we embarrass all those who, freely and candidly, shared their knowledge of the North Sea with us, and all those who made us welcome and succeeded in charming us but told us very little or nothing at all. To all of them, yet in different degrees, our warm thanks.

Much information is required for the preparation of studies such as the present one, even though the final product rarely makes explicit use of all the material collected. In seeking this information the researcher expects to be told that some of it is confidential and that certain facts and figures are commercial secrets. We have respected confidentiality in all cases and we have kept clear of areas legitimately considered to be privileged. We observed however that institutions draw the line on confidentiality in very different places. It is always amusing to find that certain facts which one respondent refuses to disclose are often well known to others outside his own institution, or that figures which some guard jealously can be easily compiled by indirect methods in painstaking but fruitful ways. There is not much we legitimately needed to know which we did not obtain in the end through hard work and a spot of good luck.

READER'S GUIDE

The results of our research on the market for North Sea crude oil are presented in this book and its annexes. All the relevant aspects of the story are told in the main body of the book. The annexes present, in a succinct form, the econometric work that supports the various analytical arguments made in the book. Finally, a statistical appendix provides a selection of relevant data on the North Sea.

This book attempts to describe as accurately as possible the important factual aspects of the North Sea market; to present all that was discovered in the course of our research and that is both new and relevant to an understanding of our subject; and finally to analyse the economic behaviour and the performance of the North Sea in relation to the world petroleum market.

Descriptive Material

The physical characteristics of North Sea crudes and blends and the relevant statistical features of the fields that produce them are presented in Chapter 2. The pattern of international trade in North Sea crude oil is described in Chapter 3. Facts about oil fiscal regimes in the UK and Norway are summarized for the benefit of the interested reader at the beginning of Chapters 8 and 10 respectively. We hope that these descriptions of complex tax systems are informative and accurate; they should not be taken however as authoritative statements on issues that may occasionally puzzle even experienced professionals with considerable legal and fiscal expertise. Those who know everything about the North Sea can skip these sections without losing on substance.

Another set of facts relate to the structure and modes of operation of North Sea markets. On structure we have been concerned with such basic questions as who are the primary suppliers and the main end-use buyers of North Sea crudes; what is their share of the total supplies and total end-use purchases; what is the degree of market concentration? We have attempted to draw a quantitative picture for 1984 of primary flows of North Sea oil from producers to refiners. The results for the North Sea as a whole are given in some detail in Chapter 3.

The same questions were asked about the spot/forward market for North Sea crudes (particularly Brent). On the basis of data obtained from *Petroleum Argus* and extensive interviews of oil companies we were able to discover relevant facts about the volume of activity, the growth of the market, the identity of participants and their shares of the total number of deals. The picture drawn is extremely up to date since the information put at our disposal covers trading in the Brent blend of crude until the end of September 1985. The results of this research are presented in Chapter 13.

The methods and procedures of trading North Sea crudes (Brent) in the forward market, and the informal, yet fairly precise, regulations of that market are explained in detail in Chapter 12. This chapter defines the terms of the paper trade in Brent and explains how 'daisy chains' are formed and how the market is made to clear when the 'month' traded forward becomes operational.

Analysis

The analytical questions investigated in this study relate to the issues listed below:

— The factors influencing production decisions in the North Sea, particularly the hypothesis that economic choice is involved, are examined in Chapter 5. Government policies relating to output maximization in the North Sea are discussed in Chapter 6 (UK) and Chapter 10 (Norway).

— The meaning and role of tax spinning by integrated oil companies with producing interests in UKCS is the subject of the final part of Chapter 8. (Interesting facts about the tax-spinning behaviour of different companies and changes in this behaviour over time may be found in sections 3.10 and 8.7.)

— The operation of the Brent market is analysed in terms of the motives and actions of hedgers and speculators in Chapter 13.

— The behaviour of spot prices, their trends and variability, the behaviour of price differentials, and the relationships between North Sea crude prices and those of other world crudes are the subject of Chapter 14. The question of whether forward prices are good predictors of the future spot price, which is fundamentally that of the economic meaning of forward trading, is discussed in section 13.6.

— Finally a broad range of questions relating to the role and influence of the North Sea in the world petroleum market are raised in Chapter 15. The role of the North Sea seems to be related to its size and centrality, to the barometric function of its prices, and to the impact of volatility on a segmented market in which OPEC has attempted to achieve a measure of stability through output adjustments.

The Past and the Future

The study examines certain historical developments (for example the origins of the Brent market in Chapter 11, the role of BNOC in Chapter 7 and of Statoil in Chapter 9) and also looks into the future. The effects of the abolition of BNOC are assessed in Chapter 3, where we examine the possible changes in the supply structure after the abolition; in Chapter 8, where we speculate on whether and in what new forms tax spinning will continue after BNOC; and in Chapter 15 where we briefly touch upon the impact of the abolition on the stability of the world petroleum market. Finally, our views about the future of forward trading are also stated in Chapter 15.

ABBREVIATIONS

API	American Petroleum Institute
APRT	Advance Petroleum Revenue Tax (UK)
ARA	Amsterdam, Rotterdam, Antwerp area
b/d	barrels per day
BGC	British Gas Corporation
BNOC	British National Oil Corporation
BP	British Petroleum
BRC	Belgian Refining Corporation
CEPSA	Cía Española de Petróleos SA (Spain)
CFP	Compagnie Française des Pétroles
c.i.f.	cost, insurance and freight
CPE	Centrally Planned Economy
CPL	Century Power and Light
CT	Corporation Tax
DNO	Det Norske Oljeselskap
EEC	European Economic Community
EMP	Empresa Nacional del Petróleo (Portugal)
f.o.b.	free on board
ICI	Imperial Chemical Industries plc
IEA	International Energy Agency (OECD)
INPC	Irish National Petroleum Corporation Ltd
IPE	International Petroleum Exchange (London)
ITO	International Thomson Organisation
KPC	Kuwait Petroleum Corporation
Kr	Norwegian krone
mb	million barrels
mb/d	million barrels per day
mt	million tonnes
mtoe	million tonnes of oil equivalent
n/a	not applicable
n.a.	not available
NCW	Non-Communist World
NGLs	Natural Gas Liquids
NOCO	Norwegian Oil Consortium AS & Co
Nymex	New York Mercantile Exchange
OECD	Organisation for Economic Co-operation and Development
OPA	Oil Pipelines Agency
OPEC	Organization of the Petroleum Exporting Countries

OTO	Oil Taxation Office
PIW	*Petroleum Intellegence Weekly*
PPB	Petroleum Price Board (Norway)
PRT	Petroleum Revenue Tax (UK)
PTA	Petroleum Taxation Act, 1975 (Norway)
RTZ	Rio Tinto-Zinc
SITCO	Shell International Trading Company
SPD	Supplementary Petroleum Duty (UK)
SPR	Strategic Petroleum Reserve (USA)
ST	Special Tax (Norway)
SUKO	Shell UK Oil
TWO	Transworld Oil
UAE	United Arab Emirates
UKCS	United Kingdom Continental Shelf
UOP	Universal Oil Products
URBK	Union Rheinische Braunkohlen Kraftstoff AG
VLCC	Very Large Crude Carrier
WM	Wood, Mackenzie & Co.
WTI	West Texas Intermediate

GLOSSARY

API Gravity	:	a scale adopted by the American Petroleum Institute for expressing the specific gravity of crude oils.
Cetane index	:	a measure of the combustion properties of diesel fuel (analogous to the octane number of gasoline) based on the gravity and volatility of the fuel. In general, a figure of 50 or above is satisfactory.
Cracker feedstock	:	oil product used as an input to cat-cracking, a particular type of upgrading (see below).
Distillation yields	:	volumes of different oil products obtained by the distillation of a barrel of crude oil.
Downstream	:	the refining and products distribution end of the petroleum industry.
Enhanced recovery	:	techniques involving the injection of chemicals into an oil reservoir, used to increase the recovery of oil in place.
Lube properties	:	properties of a particular crude that are relevant to the manufacture of lubricants.
Netback	:	method of converting the refined value of a barrel of crude oil into a price that can be compared with that of the crude oil at the point of loading by subtracting processing, transport and miscellaneous handling costs.
North Sea	:	refers exclusively to the UKCS and Norwegian Continental Shelf.
Paper barrels	:	sale/purchase *contracts* for crude oil.
Paraffinicity of vacuum distillates	:	Heavy vacuum distillates have several uses including further processing as feedstock for cracking. Paraffinic/waxy feed-stocks are generally preferred for cracking.
P/O/N/A	:	Paraffins/Olefins/Napthenes/Aromatics.
Product slate	:	typical refinery yield pattern of crude oil.
Sour crude	:	crude oil with high sulphur content.
SPAR system	:	offshore loading system at Brent field for Brent crude oil.
Sweet crude	:	crude oil with low sulphur content.
In tank ARA	:	in storage at Antwerp, Rotterdam, Amsterdam.
Tax spinning	:	refers to arm's length sales by integrated oil companies of that part of their UKCS output that they require for their own refineries or for their subsidiaries. These sales are made in order to reduce their overall fiscal liability; and the amount of crude oil required is then purchased at arm's length either in the North Sea or on other markets. Tax

		spinning replaces an internal transfer of crude oil by two arm's length transactions, a sale and a purchase.
Topping (in refining)	:	distillation process which removes the volatile fractions of a crude.
UOP K-factor	:	an index used for classifying crude oils according to their physical and chemical characteristics. It was originally based on average boiling point and specific gravity but has since been related to other characteristics of crudes. Amongst other things it provides an approximate guide to the yield of distillates from catalytic cracking of vacuum gas oil.
Upgrading (in refining)	:	refining process (separate from distillation) which improves the product yield of crude oil.
Upstream	:	the crude oil end of the petroleum industry.
Vacuum gas oil yield	:	volume of gas oil obtained by vacuum distillation.
Wet barrels	:	physical crude oil.

A SUMMING UP

The role of the North Sea in the world petroleum market is attributable to both size and function. The North Sea does not command a large share of world oil reserves; and though it ranks third in the international league of oil producers (1985), some way after the USSR and the USA, its share of world output is small. Yet North Sea production has been growing rapidly in a period of stagnant energy demand – a reflection of the output maximization strategy pursued by governments and oil companies alike. An extremely high proportion of this production is traded in the open market. Furthermore, physical trading of some North Sea crudes (mainly a particular blend of crudes known as Brent) is usually multiplied 10–15 times (and in some circumstances by much more) by paper trading in the forward market.

The North Sea occupies a central place in the world petroleum market because: (a) UKCS and Norwegian crudes have penetrated two major consuming areas – North West Europe and the USA; and (b) spot and forward prices, at least for Brent, are fairly transparent and looked upon as having barometric significance. North Sea blends and crudes are very similar in character. They are highly tradable and widely accepted by refiners in Europe and North America, and they are close substitutes for main world crudes such as West Texas Intermediate (WTI), Arabian Light, Bonny, Forcados and other North African crudes, to mention only the most important. Recently, most crude oil prices have tended to converge towards those of this central cluster of crudes, and this has increased the leverage of North Sea crudes on the whole price structure.

The high market profile acquired by the North Sea is largely due to the development of spot and forward trading in Brent. The Brent market owes its beginnings and subsequent expansion to a host of factors, among which 'tax spinning', arm's length sales undertaken by integrated companies in the UK for fiscal reasons, played a major part. Tax spinning involves financial gains under any conceivable set of circumstances provided that the arm's length price of a transaction is lower than the fiscal valuation price of oil transferred

xvii

internally by an integrated company. Though tax spinning is profitable for the company, it can have wider implications for the stability and behaviour of markets. We found that the extent of tax spinning varies from one major oil company to another, and that there are significant differences in behaviour over time.

The Brent market experienced a sudden and significant increase in activity in the period starting in mid-1984. At that time the spot price of North Sea crudes fell, inducing increased tax spinning by all the majors, and forcing BNOC to increase its recourse to the spot market. There was no growth in Brent production at the time, but there was a rise in the supply of Brent to the spot market because of the change in behaviour of BNOC and the majors. The growth in activity in the Brent market was most probably a consequence of the initial spot price decline. We do not know whether or not this expansion pushed prices further down; what is certain is that it was associated with increased price volatility.

The composition and structure of the Brent market are examined in detail. Though the number of participants in 1980–85 exceeded 110, we found that the number of continually active players was of the order of 30–35, and that the 10–15 top participants accounted for most of the activity. The top five participants comprised four oil traders and one major oil company. Out of the top ten participants, only three had primary access to the Brent blend.

In early 1985, activity in the Brent market may have reached, or even exceeded, its sustainable upper limit. The number of links in the average daisy chain increased from 10–11 to 15–16 (the average understates the length of a typical chain because it includes spot deals, for which there is no chain); at the same time the forwardness of Brent deals declined from an average of 1.5 months to 1.0. Short trading and long chains hinder the smooth clearing of the market when the traded month becomes operational. A crisis occurred in February 1986 which brought Brent trading to a standstill for a short while.

The Brent market has shown signs of strain since early 1985 and the question of greater regulation has became topical since the February 1986 crisis. This study attempts to explain these problems in terms of an equilibrium model of the number of deals in a forward market. It specifies the roles of hedgers and speculators (these terms being defined narrowly in a strict economic sense) and argues that speculators make their profits by sharing the premium that hedgers, as it were, are prepared to pay in order to buy certainty. There is no business for speculators in the absence of hedgers; and the activity of speculators who trade in paper barrels, and hence the number of forward deals, are limited in the long run by the extent of hedging.

The importance of Brent for the world petroleum market lies in the transmission of fairly (though far from perfectly) transparent price signals. Price behaviour is therefore a major issue for analysis. In this context, it is useful to recall that the emphasis on Brent does not imply that the rest of the North Sea is irrelevant, but rather that Brent is representative.

We found that spot price movements of world crudes are highly correlated, but that the correlation is stronger between North Sea prices and those of crudes traded in Europe than between spot Brent and spot WTI prices. However, the variability of spot prices differs markedly for different crudes. Volatility is highest for WTI and North Sea crudes and lowest for Middle East varieties such as Arabian Light and Dubai. Price volatility is greater in markets where output is not regulated than in the OPEC region, where an official price structure is sustained by output regulation acting as a stabilizer.

To complement the analysis of spot price movements we studied the behaviour of crude price differentials. Two regularities in the pattern of behaviour were observed: convergence over time and considerable short-term variability in the mean value of price differentials. We argue that this variability is both a consequence of trading and an inducement to trade continually and flexibly with minimum hindrance from the commitments imposed by term contracts. Oil companies need not ask themselves which of spot trading or the volatility of price differentials is cause or effect; the perception of volatility is sufficient motivation for spot trading.

The description of trends and fluctuations, which shows high degrees of correlation between long-term movements of both prices and price differentials, does not answer all the questions about the factors which influence price formation.

One aspect of this issue is the degree of market power wielded by oil companies, another is the role of OPEC. We have examined the structure of the North Sea market and measured degrees of concentration on both the supply and the end-user sides. Until the aboliton of BNOC, the supply side was dominated by BNOC itself, a few majors and Statoil. The end-user side, at least in Europe, was and will continue to be dominated by the majors.

BNOC, however, did not derive monopoly power from its large market share; it was sometimes perceived as a distress seller. Paradoxically the abolition of BNOC, which will reduce the apparent degree of market concentration on the supply side of the market, may in fact increase the relative power of the majors.

Market power is relevant to the debate on tax spinning. When the UK tax authorities valued internal transfers of oil at the BNOC price, the tax gain from spinning increased with the gap between the BNOC term price and the spot price. If integrated companies had power in both the oil products and the crude oil markets, they would benefit from a strategy that depressed crude prices and sustained those of oil products. It is certain, however, that the majors do not in fact have a strong monopolistic influence in oil products markets, and therefore that the strategy described could not be pursued.

There is also concentration in the Brent market, where BP carries visible influence (shared to some extent with a number of oil trading firms) and where Shell discreetly plays the role of oil supplier of last resort.

On these issues, we conclude that the power of major oil companies cannot influence long-term price trends. This does not mean that companies will not intervene on the market to achieve short-term economic (or, at the behest of governments, public policy) objectives. Temporary departures from the output maximization programmes are possible and have occurred in recent years. Furthermore, the Brent market presents opportunities for influencing short-period price movements by a variety of devices available to leading participants. These types of intervention may have important short-term expectational effects but do not affect the fundamentals of the market.

In contrast, OPEC has exercised a major influence on oil prices for a period of years. The OPEC price structure provides a reference point around which spot price movements in the North Sea and elsewhere take place. This reference point is effective whenever OPEC finds itself able to absorb oil demand shifts by a pure quantity adjustment; and it loses its significance whenever OPEC members try to moderate the quantity adjustment by informal price changes (discounts, counter trade, etc.).

In the slack conditions of the oil market, which have prevailed since 1980, OPEC has become vulnerable to the impact of fluctuations in a barometric price such as the North Sea. Between 1981 and 1985 OPEC attempted to stabilize oil prices by absorbing changes in oil demand by quantity adjustments. This attempt was made that much more difficult by the fluctuations of North Sea (and US) prices. OPEC members' reaction to short-period price volatility was asymmetrical. The short-period rise was always dismissed as ephemeral (because of the perception of a declining trend) and only elicited a quantity adjustment. The short-period price fall was often taken as confirmation of the trend, but gave an exaggerated impression of the underlying decline. Some OPEC members then reacted by lowering their prices.

The greater the variability of a barometric spot price – and Brent, like WTI, is very volatile – the greater is its impact on the OPEC price structure because the ups and downs are that much more frequent and that much more sharp. In the end (late 1985), OPEC gave up the attempt to 'defend' prices by passive quantity adjustments and the oil price crisis of 1986 was brought about.

CONTENTS

TABLES

FIGURES

CHAPTER 1

THE NORTH SEA IN PERSPECTIVE

1.1 Historical Development

The North Sea is an important oil and gas province which has only recently been explored and developed. The UK and Norway, which have rights in large sectors of the North Sea, and which have now become significant oil exporters, are newcomers on the world oil scene. The first exploration and production licences were granted by the UK and Norwegian Governments in 1964 and 1965 respectively. The first oil discoveries were made in 1969, and Norwegian crude oil production began in 1971 (on a very modest scale) and UKCS production in 1975.

The development of North Sea petroleum resources was fairly slow in the early stages but received a boost in the 1970s from changes in the circumstances of the international oil industry. In the late 1960s oil companies became concerned about their property rights in the OPEC region (because of nationalisations and host governments' demands for equity participation in the concessions), and they sought to diversify their sources of production. In the 1970s, hefty increases in the price of oil improved the economics of exploration and development in high-cost areas such as the North Sea.

The drilling of exploration, appraisal and development wells provides a convenient annual index of activity. Another indicator is the rig time (in rig years) spent on mobile rigs and fixed platforms. All these data are shown in Tables S.1 and S.2 in the Statistical Appendix (Annex 8).

Judging from the rig-time data, activity in the UK sector seems to have built up between 1964 and 1969. The next phases of significant growth were 1972–5 and 1979–84. In Norway, exploration activity (measured by the number of exploration wells) built up between 1966 and 1970 when fourteen wells were drilled. The next phase of increased activity lasted from 1973 to 1976 and this was followed by another burst starting in 1979–80.[1]

---

[1] For an explanation of these episodes in exploration, see Hall, S.G. and Atkinson, F., *Oil and the British Economy*, Croom Helm, London 1983.

In the UK, an early oil discovery was made in the Hewett field in November 1966, but Hewett has not been developed for oil production. After a lapse of three years Montrose was discovered in December 1969. However, production in this field did not start until June 1976. The first major discovery in the UK sector was Forties which BP found in November 1970. Brent, the second largest UK field, was discovered in July 1971. Argyll was the first field to produce oil in the UKCS (June 1975), Forties began producing in September 1975, and Brent in November 1976.

In Norway, Ekofisk was discovered in December 1969 and a very small production volume came on stream in July 1971. Ekofisk remained the only Norwegian producing field until Cod and West

Table 1.1:   UK and Norway. Crude Oil Production (Offshore). 1971–85. Various Estimates. Units as shown.

	UK			Norway		
	Brown Book (mt)	BP (mt)	WM ('000b/d)	Fact Sheet (mtoe)	BP (mt)	WM ('000b/d)
1971	–	–	–	0.3	0.3	–
1972	–	–	–	1.6	1.6	33
1973	–	–	–	1.5	1.8	32
1974	–	–	–	1.7	1.7	34
1975	1.1	1.4	23	8.7	9.3	190
1976	11.6	11.8	242	13.9	13.8	280
1977	37.3	37.5	772	13.5	13.5	280
1978	52.8	53.2	1095	16.9	17.2	358
1979	76.5	77.8	1578	18.8	18.8	407
1980	78.7	80.3	1633	24.4	25.8	528
1981	87.7	87.2	1802	23.5	24.9	507
1982	100.1	103.2	2066	24.5	24.6	531
1983	110.5	114.6	2296	30.6	30.5	658
1984	120.8	125.6	2527	n.a.	35.0	722
1985	n.a.	n.a.	2558	n.a.	n.a.	807

Note: BP data adjusted to exclude onshore production (UK)
Sources: Department of Energy, UK, *Development of the Oil and Gas Resources of the United Kingdom*, referred to as the *Brown Book*, various issues
BP Statistical Review of World Energy, various issues
Wood, Mackenzie & Co., *North Sea Report*, various issues
Royal Ministry of Petroleum and Energy, Norway, *Fact Sheet*, various issues

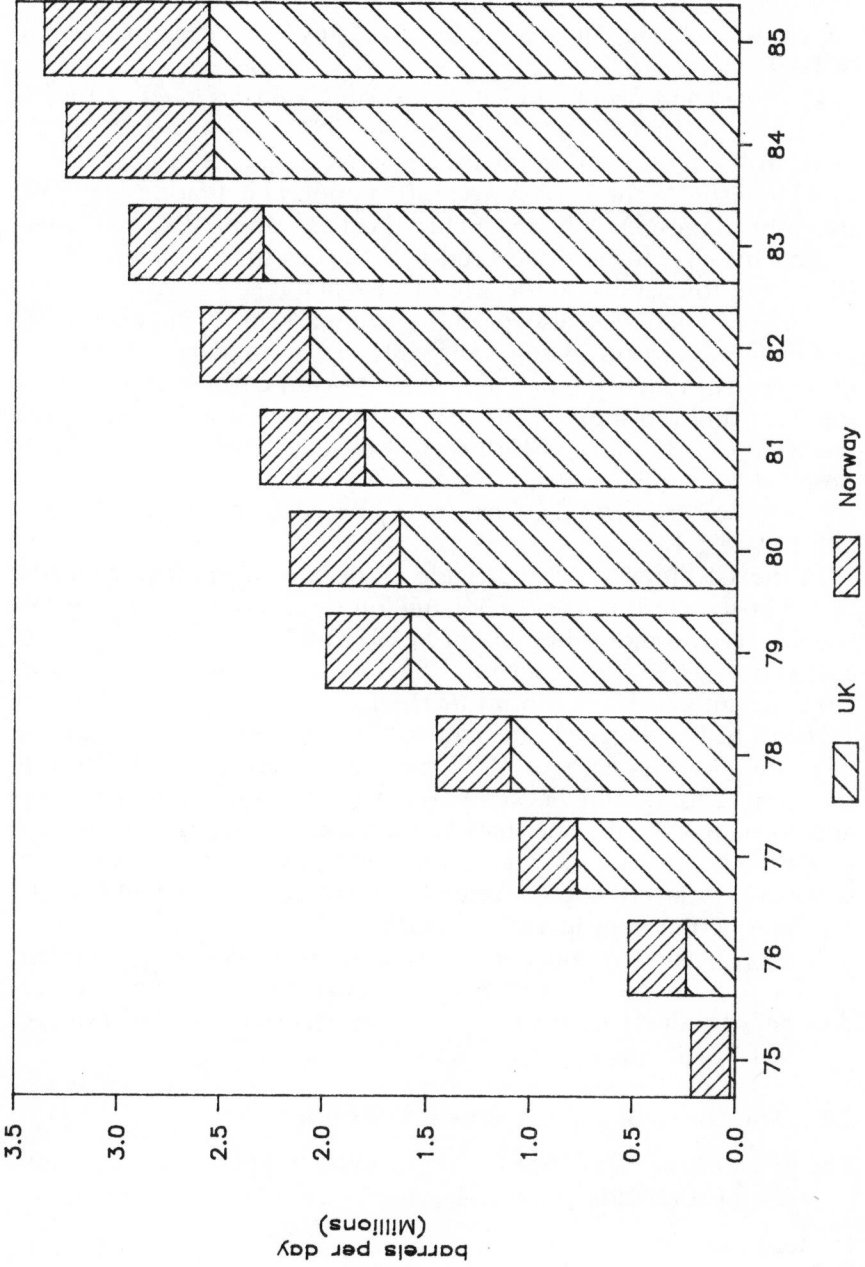

Figure 1.1 UK and Norway Crude Oil Production (Offshore)

Ekofisk entered into production in 1977. The largest Norwegian oilfield is Statfjord. Discovered in 1974, this field began producing in December 1979.

Discovery and production start-up dates are presented in Tables S.3 and S.4 (for fields producing oil in 1985). These include thirty fields in the UKCS and ten fields in the Norwegian sector (seven of which constitute the Ekofisk area). It is noticeable that three out of the four largest North Sea fields (Forties, Brent, Ekofisk) were among the first to be discovered and to start up production. (See Chapter 2 for details on the size of fields.)

Tables S.3 and S.4 also permit a comparison of discovery and production start-up dates. The difference may be called the development time-lag. For the UK the shortest development lag was one year and ten months (Innes); for Norway it was one year and seven months (Ekofisk). More typical development lags are of the order of four to six years.

North Sea crude oil production developed as shown in Table 1.1 and Figure 1.1.

In the UK, total stabilized crude oil output from offshore fields began at the modest level of 1.1 million tonnes (mt) in 1975 but it rapidly grew to more than 50 mt in 1978 and more than 100 mt in 1982. It reached 120.8 mt in 1984. In Norway output increased from 0.3 mt in 1971 to 35 mt in 1984.

North Sea oil entered international trade soon after the start-up of production. As early as 1977 the UK was an exporter of some 15.6 mt. This should be compared with a production figure of approximately 37 mt. The share of gross exports in crude oil output had already reached the high level of 42 per cent two years after production started, and continued to increase in subsequent years, reaching 60 per cent in 1983 and 1984.

Norway began to export oil in 1972, almost as soon as production had begun. In 1976 these exports represented 89 per cent of Norway's production. Since 1981 the export share has been of the order of 82–85 per cent.

1.2   The North Sea in the World Oil Industry

The place of the North Sea in the world oil industry can be gauged in terms of such indicators as:

(a) Reserves
(b) Production
(c) Investments
(d) Exports

(a) Reserves. The 1985 *Brown Book* puts UK proven oil reserves at 1500 mt (end 1984), and proven plus probable reserves at 2000 mt. An estimate of 650 mt of possible reserves brings the aggregate to 2650 mt. Following the *Oil & Gas Journal*, the *BP Statistical Review of World Energy* records 1800 mt of proven reserves for the UK.

Norway's Petroleum Directorate puts proven oil reserves at about 1000 mt (end 1983) but recognises that this estimate is lower than that of the licensees.[2] BP reports 1100 mt (end 1984).

Comparisons with reserve estimates in other parts of the world should be treated with great care. Their only purpose is to give a rough idea of the importance of the North Sea relative to other oil provinces. These comparisons are made in Table 1.2.

The reserves criterion, taken at face value, suggests that the North Sea is but a small part of the world oil industry. Either OPEC or Middle East reserves dwarf the North Sea into insignificance.[3]

However, the reserves/production ratios are more favourable for the North Sea (UK was 14.4 years and Norway 31.9 years at the end of 1984) than for the USA (8.9 years) or some of the new oil-exporting countries such as Egypt (9.8 years) and Dubai (11.0 years). Norway's reserves/production ratio is higher than that of several OPEC or other Middle East countries such as Ecuador, Oman, Qatar, Syria, Algeria, Indonesia, etc.

There seems to be a consensus of opinion about an early peak in North Sea oil production. In March 1985 the UK Minister of State for Energy told Parliament that crude oil and natural gas liquids (NGLs) output for the UK was forecast as follows (million tonnes)[4]:

1985	120–135
1986	110–130
1987	95–125
1988	85–120
1989	80–115

[2] Royal Ministry of Petroleum and Energy, *Fact Sheet*, 1984:1, p10.

[3] Reserve estimates are notoriously uncertain. The various definitions used (proven, probable, possible) involve criteria and probability assessments that leave much room for individual judgement. The issue of North Sea hydrocarbon reserves has been surrounded by controversy, especially in the late 1970s. For example, Peter Odell has consistently argued that the estimates of the Department of Energy and those of the oil industry grossly understate the size of North Sea reserves. The diversity of views in the early debate is best illustrated by a comparison of Robinson, C. and Morgan J., *North Sea Oil in the Future*, Kogan Page, London, 1978, with Odell, P.R. and Rosing, K.E., *The North Sea Oil Province* and *The Future of Oil*, Kogan Page, London, 1975 and 1980 respectively. On this issue , see also Hall S.G. and Atkinson, F., *op.cit.*

[4] See *Brown Book*, 1985, p81.

Table 1.2:   UK and Norway. Proven Oil Reserves Compared with other Areas.
End 1984. Thousand Million Tonnes and Percentages.

Area	'000 mt	%
UK	1.8	1.9
Norway	1.1	1.2
Total Western Europe	3.3	3.5
USA	4.4	4.9
USSR	8.6	8.8
China	2.6	2.7
Middle East	54.2	56.4
Latin America	11.7	11.8
Africa	7.5	8.0
Asia and Australasia	2.5	2.5
Total World	96.1	100.0
of which OPEC	64.8	67.2

Source: *BP Statistical Review of World Energy*, 1985

This implies a UK peak in 1985. However, some experts believe that the peak will not be reached until 1986 or even 1987. Norway's oil production is expected to peak at a later date, probably in the first years of the 1990s.

Projected production profiles for the North Sea as a whole provide very different pictures depending on estimates of ultimately recoverable resources and assumptions about production policies. The production decline could begin as early as 1996 or as late as 2013, according to various projections made by the US Department of Energy in 1983.[5] These projections expect the North Sea to be depleted of oil some time between 2020 and 2040.

In short, North Sea oil reserves are small compared with reserves in the Middle East, Mexico or the USSR. They do not compare unfavourably with Nigeria, Libya, Venezuela or even the USA. For the latter, the resource base is larger than for the North Sea but depletion rates are much faster.

(b) Production. The oil production share of the North Sea is larger than the proportion of world reserves that the area commands. In 1984, UKCS production was about 2.5 mb/d and Norway's 0.7 mb/d. These two countries' combined share of total world oil output was 5.7 per cent. Their share of oil production in the non-Communist world was 7.5 per cent.

[5] Energy Information Administration, US Department of Energy, *The Petroleum Resources of the North Sea*, 1983, pp52–56.

Table 1.3:   UK and Norway. Crude Oil Production Compared with other
Countries. 1983 and 1984. Million Barrels per Day.

	1983		1984	
	mb/d	%	*mb/d*	%
UK	2.36	4.2	2.58	4.5
Norway	0.63	1.1	0.71	1.2
Total Western Europe	3.49	6.2	3.80	6.5
USA	10.22	17.6	10.39	17.5
USSR	12.52	22.4	12.42	21.7
China	2.14	3.8	2.30	4.1
Middle East	12.01	21.4	11.74	20.3
Latin America	6.43	11.7	6.71	11.9
Africa	4.77	8.4	5.19	8.9
Asia and Australasia	2.91	5.2	3.32	5.8
Total World	56.40	100.0	57.80	100.0
of which NCW	41.34	73.1	42.69	73.5
of which OPEC	18.28	32.5	18.35	31.8

Source : *BP Statistical Review of World Energy*

(c) Investments. The North Sea has absorbed huge sums in the form
of exploration expenditure and development investments. Wood
Mackenzie have recently provided estimates of the total development costs of forty-three fields in the UK and Norwegian sectors
(Ekofisk counting as one field). Unfortunately, exploration expenditures are not included. Data from the *Brown Book* for 1976–84
suggest that exploration outlays correspond to some 25 per cent of
development costs.

Applying this mark-up suggests that expenditure on exploration
and development between 1976 and 1984 lay in the range of $4–9
billion per year. These figures may be compared with Chase
Manhattan's estimates of capital and exploration expenditures in
the non-Communist world.[6] According to Chase, these expenditures rose from $23.9 billion in 1976 to $88.8 billion in 1982. The
North Sea share can be put at 18 per cent in 1976, but this
proportion declined rapidly in subsequent years, falling to 10 per
cent in 1982.

[6] Energy Economics Group, Chase Manhattan Bank, *Capital Investments in the World
Petroleum Industry*, 1982, p17.

The North Sea investment share appears to be very large in the world outside the USA. This is because a big proportion of the world expenditures on exploration and development of oil and gas resources is attributable to the USA (about 55 per cent between 1976 and 1982). In 1976, the North Sea accounted for 40 per cent of exploration and development expenditure in the world outside the USA while the Middle East share was just over 13 per cent. In 1982 the North Sea and Middle East shares were 24 and 9.7 per cent respectively.

(d) Exports. The significance of the North Sea is at its highest when the assessment is made in terms of crude oil exports. In 1984, total world oil exports were of the order of 21.21 mb/d and world exports excluding those of the Communist countries were about 18.07 mb/d. The gross exports of the UK and Norway were 2.17 mb/d, slightly more than 10 per cent of world and about 12 per cent of non-Communist world exports. These figures should be compared with a 3.1 per cent share of world reserves and a 5.7 per cent share of world production.

Table 1.4:   UK and Norway. Crude Oil Exports Compared with Major Oil-exporting Countries. 1984. Million Barrels per Day and Percentages.

	mb/d	%
UK	1.55	7.3
Norway	0.62	2.9
USA	0.18	0.8
USSR	2.68	12.6
China	0.41	1.9
Middle East	8.01	37.8
Latin America	2.81	13.2
Africa	2.97	14.0
Asia and Australasia	1.54	7.3
Total World	21.21	100.0
of which NCW	18.07	85.2
of which OPEC	11.84	55.8

Source: OPEC, *Annual Statistical Bulletin*, 1984.

North Sea crudes are important in the context of the international oil industry for two main reasons:

— A very large proportion of North Sea production is traded internationally.

— The North Sea share of crude in international trade is a significant proportion of about 10 per cent.

1.3 The Significance of the Markets for North Sea Crudes

The importance of markets for North Sea crudes is both qualitative and quantitative. We have seen that the North Sea occupies a larger place in international crude trade than its shares of world production and reserves would have led us to expect.

This place was acquired for a variety of reasons. Norway is a very small consumer of oil and a very large proportion of its petroleum production is an exportable surplus. The UK is a much larger consumer but it still exports an amount well in excess of its true surplus (measured as the difference between production and domestic consumption). This is due partly to the structure of the refinery capital stock, some of which is better geared to using foreign crudes or feedstocks, and partly to commercial factors, since oil companies are always looking for the most economical package of crude oil inputs for their refineries and this may well entail selling North Sea crudes and buying some other varieties.

The North Sea enjoys a tremendous locational advantage. Export terminals are within one to two days' sailing of the refining centres of North West Europe. The North Sea is a major producing area situated close to one of the three major consuming areas of the world (Western Europe, the other two being the USA and Japan).

The emergence of the North Sea as an oil-producing region within Europe has caused some changes in the structure of the world oil industry:

— The import dependence of Western Europe, and to some extent of the USA, on imports from the OPEC region has declined.

— The North Sea has enabled refiners in North West Europe to adopt more flexible supply policies. Commercial inventory requirements have declined because North Sea supplies are available almost immediately, being on short haul. In the 1970s supplies involved a 30- to 40-day voyage from the Gulf.

— The development of spot and forward markets in North Sea crudes has made a major contribution to this flexibility. End-users can obtain supplies on a day-to-day basis if they are willing to accept the risks of unexpected spot price changes. But they can also hedge on the forward market or engage in parallel speculation.

The rapid development of spot and forward markets has given the North Sea crude markets a barometric role. Price movements of Brent are carefully watched by all traders as are the spot and futures prices of WTI in Texas and New York, the spot prices of products in the main refining centres and certain 'critical' crudes such as Urals in Europe and Dubai in the Middle East.

In the past two or three years, North Sea production and exports have increased at the rate of 10 per cent per annum. The North Sea is thus acquiring a great significance at the incremental margin of oil trade. These high rates of growth are obtaining in a period of stagnant world demand and declining OPEC output. This relentless surge in output and exports is eroding both OPEC's share of the world petroleum market and its ability to defend its official oil price structure.

PART I

MARKET STRUCTURE, TRADE AND PRODUCTION

CHAPTER 2

NORTH SEA CRUDES AND OILFIELDS

2.1 Introduction

A market is a set of economic relationships between the sellers and buyers of some commodity, good or service. A market is initially identified first by its location and, more importantly, by the item traded. It is therefore natural to begin this study of the market for North Sea crudes with some remarks on the location and with a more substantial description of the characteristics of the crude varieties involved in North Sea transactions.

The location of a market may be defined in different ways: (a) the place where deals are made, and (b) the geographical area in which the commodity traded in these deals moves. Thanks to the development of modern communications, deals can be effected anywhere and do not generally require the physical presence of buyers, sellers or intermediaries on the same spot. In that sense, it is correct to say that the market for North Sea crudes has no specific location. Deals can be made virtually anywhere in the world. There is no institutionalized Exchange with a floor for spot and forward transactions in North Sea crudes, and until 1985 (save for an earlier but very short period) there was no futures market. In practice, however, a large proportion of arm's length deals are done in the UK, Norway and the USA, and London is generally considered as the privileged locus of this trade.

Another geographical definition of a market refers to the area in which the commodity moves. We shall see in Chapter 4 on international trade in North Sea crudes that virtually the whole output volume moves in North West Europe and the USA. The physical market in North Sea crudes has very discernible boundaries.

We may now turn to the commodity traded. North Sea crudes can be characterized in very broad terms by the following features:

(a) Location: They are (North) Atlantic Basin crudes with major outlets in North West Europe and North America.
(b) Gravity: North Sea crudes are generally light or extra-light in terms of API gravity, but there are exceptions.

(c) Sulphur: Sulphur content is generally low but there are exceptions.

This broad and generally accepted description is not very accurate. North Sea crudes display much wider variations in characteristics than is usually acknowledged. This is partly a result of the large number of producing fields from which these crudes are extracted. By end 1985 there were twenty-eight producing fields in the UK, eight producing fields in Norway and two fields straddling across the UK and Norwegian sectors.

The range of characteristics may be summarized as follows:

(a) API gravity range: 29° (Claymore) to 44° (Albuskjell and Cod).
(b) Sulphur content: 0.2 per cent (Ekofisk, Duncan, Montrose) to 1.9 per cent (Claymore).
(c) UOP K-factor: 12.10 (Ekofisk, Brent blend) to 11.80 (Flotta).
(d) Yield of total distillate: yields at C_5–350°C range between 55 per cent of volume (Flotta) and 68.2 per cent (Ekofisk).
(e) Yield of cracker feedstock: yields at 370–520°C range between 14.6 per cent (Ekofisk) and 23.8 per cent (Ninian blend).

But many of these differences are blurred, and the range of characteristics is narrowed by the blending of North Sea crudes in the main pipeline systems. For example, Claymore is not available as such, but is part of the Flotta blend where its qualities are somewhat improved by the other crudes in the mix, namely Piper and Tartan.

We would expect North Sea crudes in the importing countries of North West Europe and North America to compete mainly with African crudes because of similar API and sulphur characteristics. They would also compete with light, sulphur-free crudes from other sources with access to the Atlantic Basin: the USSR, the Gulf, Mexico or Venezuela. In the internal US market they would also compete with similar non-exportable US crudes such as WTI.

However, technical progress has increased the flexibility of modern refineries so much that crude differentiation has lost its former importance. Indeed, this is reflected in the continual narrowing of oil price differentials on the world petroleum market. The optimization problem for a modern refinery depends more on product yields and spot product prices than on the physical characteristics of this or that crude.

This means, in principle at least, that North Sea crudes can compete with a wide range of other crudes used in modern upgraded refineries. Problems may arise, however, when specialist uses are involved, that is when a particular crude is required, say, for its lube properties, or for the production of asphalt. In addition some crudes can cause problems either because of adverse characteristics, for example Maya (Mexico) with its high metal content, or because they fail to meet local product quality specifications.

Table 2.1: UK and Norway. Producing Oilfields Ranked by Output Volumes. Annual Averages. 1984. Thousand Barrels per Day

1.	Brent (UK)	427
2.	Forties (UK)	416
3.	Statfjord (Norway)	384
4.	Ninian (UK)	237
5.	Piper (UK)	182
6.	Ekofisk (Norway)	147
7.	Fulmar (UK)	126
8.	Magnus (UK)	109
9.	Claymore (UK)	100
10.	North Cormorant (UK)	93
11.	Beryl (UK)	89
12.	S Brae (UK)	88
13.	Thistle (UK)	87
14.	Murchison (UK)	81
15.	Maureen (UK)	77
16.	Statfjord (UK)	73
17.	Dunlin (UK)	65
18.	Eldfisk (Norway)	62
19.	Valhall (Norway)	52
20.	NW Hutton (UK)	51
21.	Beatrice (UK)	46
22.	Cormorant (UK)	35
23.	Tartan (UK)	29
24.	Murchison (Norway)	27
25.	Hutton (UK)	25
26.	Heather (UK)	24
27.	Buchan (UK)	21
28.	Tor (Norway)	19
29.	Montrose (UK)	16
30.	Auk (UK)	12
31.	Albuskjell (Norway)	11
32.	W Ekofisk (Norway)	11
33.	Duncan (UK)	10
34.	Argyll (UK)	8
35.	Cod (Norway)	5
36.	Edda (Norway)	5
37.	Deveron (UK)	2

Source: Wood Mackenzie

2.2   Ranking of North Sea Fields and Crudes

Table 2.1 ranks the producing fields of the UKCS and Norway in terms of average 1984 daily output. Statfjord and Murchison appear twice in the table because these fields are shared (in unequal proportions) by the two countries. The ranking would be modified if these shared fields were treated as single units. In that case, the number of fields would be reduced to thirty-five, Statfjord would become the largest field with an output of 457,000 b/d, and Murchison would rank ninth with an output of 108,000 b/d.

Table 2.2:   North Sea Oilfields. Size in Terms of Peak Recovery. Barrels per Day.

Very Small (<50,000)	: Argyll, Auk, Balmoral, Buchan, Cyrus, Deveron, Duncan, Heather, Highlander, Innes, Montrose, Tartan.
Small (50,000–<100,000)	: Alwyn N, Beatrice, Clyde, Hutton, NW Hutton, Maureen, Ula, Valhall.
Medium (100,000–<250,000)	: Beryl, Brae, Claymore, Cormorant, Dunlin, Fulmar, Gullfaks, Magnus, Murchison, Oseberg, Thistle.
Large (250,000–<500,000)	: Brent, Ekofisk, Ninian, Piper.
Very Large (500,000+)	: Forties, Statfjord.

But oilfields should be ranked according to criteria other than production in a given year. The output criterion is misleading because the common year chosen for comparison purposes corresponds to different points of the producing lives of the various fields. A large field caught at the beginning of its life would appear smaller than, say, a medium field near its peak. Table 2.2 ranks producing fields and fields currently under development or expected to come on stream in the next few years according to their expected peak recovery volumes. It shows that there are no large or very large fields, other than currently producing fields, expected to come into production in the next few years. There are however two medium-sized fields (Gullfaks and Oseberg), both in the Norwegian sector, which are not yet producing but which will add considerable amounts to availabilities in the late 1980s/early 1990s.

Another way of assessing the supply pattern in the North Sea is to examine the expected peak dates of various fields. This is shown in Table 2.3. If the prognoses are right, which they seldom are because of an inherent pessimistic tendency in supply forecasts, UKCS crude production should peak in 1985 or at the latest in 1986. This is because all big fields other than Statfjord peaked well before that date. Medium-sized fields, together with Statfjord, carry the thrust between 1986 and the early 1990s. These are Cormorant and Brae (both of which are now partly or fully in production) and Gullfaks and Oseberg (which are not yet on stream).

Table 2.3: North Sea Fields by Size and Dates of Peak Production

Peak Year	Very Small Fields	Small Fields	Medium Fields	Large Fields	Very Large Fields
1977	Argyll Auk				
1978					
1979	Montrose			Piper	
1980				Ekofisk	Forties
1981					
1982	Heather		Thistle	Ninian	
1983	Buchan		Murchison		
1984	Tartan Duncan	Maureen	Claymore Dunlin Fulmar	Brent	
1985	Deveron Innes	NW Hutton Beatrice	Magnus Beryl		
1986	Highlander	Hutton	Cormorant		
1987		Valhall			
1988	Cyrus Balmoral	Clyde			Statfjord
1989		Ula			
1990		Alwyn N	Gullfaks Brae		
1991			Oseberg		

Companies with producing interests in the North Sea may invest in enhanced recovery to obtain additional volumes from their fields. This could change the long-term supply picture to some extent. Whether these decisions will be made, and the scale of the investments undertaken if they are, will depend on three key factors:

— the costs of investments
— the expected price of oil
— the relevant tax concessions

On the latter, a recent study by Professor Kemp indicated that investments in enhanced recovery would become profitable in the UK given a modicum of tax incentives.[1] The Chancellor did not seem to have had time to examine this view before the 1985 Budget which did not mention the subject. Our view is that the UK will encourage these investments in the future with the appropriate incentives, but that the changes in the fiscal legislation will not be forthcoming until UKCS output begins to decline.

Table 2.4:   UK and Norway. Oilfields Ranked by API Gravity of Crudes. 1984. °API. Output in Thousand Barrels per Day.

°API	Fields	Output
43+ to 44	Albuskjell, Cod	16
41+ to 43	W Ekofisk	11
39+ to 41	Fulmar, Montrose, Tor	161
38+ to 39	Argyll, Brent, Magnus, Statfjord	1001
37+ to 38	Auk, Beatrice, Deveron, Edda, Eldfisk, Murchison, Tartan, Thistle	351
36+ to 37	Beryl, Brae, Forties, NW Hutton, Piper	826
34+ to 36	Cormorant, N Cormorant, Duncan, Dunlin, Ekofisk, Heather, Maureen, Ninian, Valhall	740
32+ to 34	Buchan	21
30+ to 31	Hutton	25
29	Claymore	100

Sources: Wood Mackenzie; *Oil & Gas Journal*; Lovegrove, Guide to *Britain's North Sea Oil and Gas*.

Having ranked North Sea oilfields by size, using a number of different criteria, we turn now to a ranking of crudes by physical characteristics. Table 2.4 ranks fields by the API gravity of their

[1] Kemp, A.G. and Rose, D., *Fiscal Aspects of Incremental Investments in the UK Continental Shelf*, University of Aberdeen North Sea Studies, Occasional Paper No.20, December 1984.

crudes. It appears that the bulk of North Sea crudes (90 per cent) have an API gravity in the range of 34–39°. The median of the distribution is at 37°API and the mode in the 38°+ to 39° bracket. As mentioned before the range is fairly wide as it extends from 29° to 44°API.

Table 2.5 shows that the bulk of North Sea crudes are very sweet as 70 per cent of 1984 output had no more than 0.35 per cent of sulphur content. This puts them in the same category as Libyan and Nigerian crudes.

Table 2.5: UK and Norway. Oilfields Ranked by Sulphur Contents of Crudes. Per cent. Output in Thousand Barrels per Day. 1984.

% Sulphur	Fields	Output
up to 0.25	Albuskjell, Argyll, Brent Cod, Duncan, Edda, Ekofisk, W Ekofisk, Eldfisk, Montrose, Tor	721
0.26 to 0.35	Beatrice, Brae, Deveron, Forties, Fulmar, NW Hutton, Magnus, Murchison, Statfjord, Thistle, Valhall	1542
0.36 to 0.45	Auk, Beryl, Dunlin, Ninian	403
0.46 to 0.65	Heather, Maureen, (Tartan)	101–130
0.66 to 0.75	Hutton, (Tartan)	25–54
0.76 to 1.00	Buchan, Cormorant, N Cormorant, Piper	331
1.8 to 1.9	Claymore	100

Sources: As in 2.4.

North Sea crudes are generally sweeter than virtually all Gulf crudes and such Atlantic Basin crudes as the Venezuelan and the Mexican. Only Algerian crudes (Saharan blend and Zarzaitine) have less sulphur than the sweetest North Sea crudes.

The median of the distribution by sulphur content (1984) is at 0.3 per cent. The mean is estimated at 0.42 per cent, and this reflects the weight of crudes with high sulphur contents at one end of the distribution. The (relatively) sour crudes are Cormorant, Piper and Claymore.

2.3 The Array of Available Blends and Crudes

North Sea crudes are not always available in single varieties for domestic use or export shipments. Many crudes find their way to the market as part of a blend. Blends have come into existence because of the structure of collecting systems. It proved convenient and economical to gather oil from a set of neighbouring fields in single pipeline collecting systems; and the solution to this transportation problem naturally gave rise to a particular blend.

Blends are what they are, given the characteristics of fields that happen to be situated within reach of an integrated pipeline collecting system. They are not the result of carefully designed mixes that aim to create a commodity with a set of attractive and consistent characteristics, but rather the by-product of a transport infrastructure.

There are four major pipeline collecting systems in the UK and one in Norway which today give rise to blends. These are:

— The Brent system at Sullom Voe, Shetland
— The Ninian system at Sullom Voe
— The Forties system at Hound Point, Firth of Forth
— The Flotta system at Scapa Flow, Orkney
— The Ekofisk system (Norway) at Teesside in the UK

(a) The Brent System. This involves Brent, Cormorant and North Cormorant, Deveron, Dunlin, Hutton, North West Hutton, Murchison (UK and Norway) and Thistle. In 1984 the combined production of these fields averaged of 893,000 b/d. The Brent crude can also be loaded off shore on the SPAR system which can handle up to 250,000 b/d in good weather. After the construction of the pipeline, the export rate of Brent through SPAR has averaged 65,000 b/d with the use of two tankers at the rate of four liftings per month. Thus Brent blend average daily volume in 1984 may well have been of the order of 830,000 barrels.

(b) The Ninian System. This includes three fields: Heather, Magnus and Ninian. Their combined output in 1984 was 370,000 b/d on average.

(c) The Forties System. This system collects crudes from three fields: Brae, Montrose and Forties. Brae may well have proved to be a nuisance, lowering the qualities of Forties. The combined 1984 output was 520,000 b/d.

(d) The Flotta System. It blends Piper, Claymore and Tartan. These three fields have one characteristic in common, namely high sulphur content (by North Sea standards). Piper and Tartan compensate to

some extent for the heavy gravity of Claymore. However, Flotta is relatively heavy and sour, and its price accordingly tends to be below that of other North Sea blends. In 1984 the combined production of the three fields was 311,000 b/d.

(e) The Ekofisk System. It blends Albuskjell, Cod, Edda, Eldfisk, Ekofisk, West Ekofisk, Tor and Valhall. The pipeline system collects from all Ekofisk fields plus Valhall and transports all the production through a main pipe to Teesside in the UK. The combined output in 1984 was 312,000 b/d.

Blends combine crudes from twenty-six fields (counting Murchison as two and the Ekofisk area as seven) into five mixes. Thus the marketed varieties of North Sea crudes (1984) consisted of five blends and ten single crudes.[2] By order of quantitative importance these were:

Blends			Single Crudes	
1.	Brent	(830–893)	1. Statfjord	457
2.	Forties	520	2. Fulmar	126
3.	Ninian	370	3. Beryl	89
4.	Ekofisk	312	4. Maureen	77
5.	Flotta	311	5. Beatrice	46
			6. Buchan	21
			7. Montrose	16
			8. Auk	12
			9. Duncan/Argyll	10/8
			also Brent	(0–63)

All single crudes, except Beatrice which has its own pipeline terminal at Nigg Bay, are loaded into tankers off shore.

2.4 Refining Characteristics of Main Blends and Crudes

The five blends plus Statfjord represented some 87.5 per cent of the North Sea oil output in 1984. We should therefore concentrate on the refining characteristics of these six varieties (plus one or two important single crudes such as Fulmar or Beryl) in order to define

[2] The higher figure for Brent includes the production from the Brent field loaded off shore by the SPAR system.

the nature of those commodities actually traded in the North Sea market.

We indicated earlier that the wide range of characteristics of individual North Sea crudes is narrowed to some extent in the blends. The varieties available to the market, though clearly distinct, are much closer to each other in their properties than the individual crudes as extracted from their respective oilfields. It should be noted, however, that the properties of a given blend do not remain constant over time. The characteristics vary because the composition of the blend changes with the vagaries of production in the constituent fields.

The properties of the five North Sea blends and of major North Sea crudes are summarized in Table 2.6. These properties are identified by particular assays made at given dates. Assays based on samples taken at different times are unlikely to produce identical results.

Crudes and blends of crudes have a variety of physical characteristics. Two crudes may be very similar in some respects and sufficiently different in others as to involve differences in costs or returns when processed by a given refiner in given market conditions.

This wide-ranging differentiation of properties (see the list of possibly relevant characteristics mentioned below) makes it difficult to identify at first sight crudes which are very close substitutes for each other, and to distinguish them from crudes belonging to other sets of close substitutes.

The problem is to group crudes in sets which display both homogeneous properties within sets and marked (i.e. statistically significant) differences between sets and which may be ranked in some ordinal way.

A possible list of relevant characteristics is:

— refining yield
— sulphur content
— wax content (paraffinicity of vacuum distillates)
— P/O/N/A percentage weights of naphtha and gasoline
— UOP K-factor
— cetane index of middle distillates

A shorthand index for yields is API gravity. But yield characteristics are more accurately measured by considering:

— yields of different cuts in topping and upgrading units
— yields of total distillate
— yields of gasoline/naphtha

Table 2.6 : UK and Norway. Refining Characteristics of Main Blends and Crudes

Characteristics	Ekofisk	Brent	Forties	Ninian	Flotta	Statfjord	Fulmar	Beryl
Crude oil:								
API Gravity	42.50	37.20	37.10	36.10	35.70	37.70	39.90	37.00
Sulphur % (weight)	0.18	0.35	0.35	0.39	1.14	0.29	0.27	0.39
Distillation % (volume):								
C_1–C_4	1.8	4.2	4.3	2.5	4.3	3.2	3.0	2.3
C_5–165°C	37.2	24.4	22.5	23.1	22.7	25.1	28.3	26.0
165–235°C	13.2	12.1	12.2	12.8	10.8	13.2	15.1	12.5
235–350°C	17.8	21.1	21.9	20.8	21.5	22.0	21.0	20.7
350+°C	30.0	38.2	39.1	40.9	40.7	36.5	32.6	38.6
Cracker Feed (370–520°C):								
% (volume)	14.60	22.50	23.50	23.80	22.30	21.30	23.10	n.a.
Sulphur % (weight)	0.23	0.40	0.42	0.52	1.66	0.40	0.40	n.a.
K-factor	12.10	12.10	11.90	12.00	11.80	12.00	12.00	n.a.

Sources : Various oil companies

Crudes also have properties relevant to their use as cracker feedstock. These are the vacuum gas oil yield (at say 370–520°C), sulphur content and K-factor.

Table 2.6 shows that two main blends and one crude – Brent, Forties and Statfjord – resemble each other very closely. Their distillation yields and, more importantly, their cracker feed characteristics are very similar. These three varieties accounted for 57 per cent of North Sea oil production in 1984. Furthermore, Forties and Brent both have good lube properties, though Brent may be superior in this respect. Ekofisk is a much lighter blend with a high yield of light distillates but its yield of cracker feed is rather low. Ninian is closer to the core varieties (Brent/Forties/Statfjord) than either Ekofisk or Flotta. The Flotta system mixes crudes with a range of API gravities thus creating a blend only slightly heavier than Ninian. The distillation and cracker feed yields are not significantly different. However, Flotta has a relatively high sulphur content compared with other North Sea varieties and this feature sharply distinguishes Flotta from Ninian, reducing its value as cracker feedstock.

2.5 North Sea Crudes and Blends Compared with World Varieties

North Sea crudes and blends compete in the Atlantic Basin (mainly the USA and North West Europe) with crudes of different origins, many of which have similar characteristics. Table 2.7 presents assays of some of these crudes. A comparison of these with North Sea blends and crudes reveals the similarity in properties of WTI and Brent blend. Saharan blend resembles Ekofisk in terms of low cracker feed yields. Bonny Light displays a distinctive pattern of characteristics, some similar to Ekofisk, while others single it out as a medium gravity crude with a high middle-distillate yield. Both Urals and Arabian Light have characteristics which lie outside the typical ranges for North Sea crudes, except for some features shared with Flotta.

A more sophisticated comparison can be made using certain statistical techniques which enable us to group crudes and crude blends in sets or clusters with similar properties within sets. Using the list of characteristics mentioned above, and drawing on work done by others, we can place the main North Sea blends and some single North Sea crudes in the following sets:

(a) *Ekofisk* with Bonny Light (Nigeria) and Saharan blend (Algeria).

Table 2.7: World Crudes. Refining Characteristics of Selected Crudes Marketed in the Atlantic Basin

Characteristics	WTI	Bonny Light	Forcados	Saharan Blend	Zarzaitine	Zueitina	Urals	Arabian Light
Crude oil:								
API Gravity	38.60	36.70	29.20	46.20	42.30	41.60	32.50	33.40
Sulphur % (weight)	0.45	0.11	0.19	0.11	0.09	0.31	1.38	1.79
Distillation % (volume):								
C_1–C_4	1.1	2.3	1.0	5.4	3.5	2.2	n.a.	1.8
C_5–165°C	23.8	21.6	15.1	35.3	28.1	28.2	23.8	20.3
165–235°C	14.8	14.5	13.1	16.0	13.8	15.2	11.2	13.6
235–350°C	22.8	29.5	34.6	20.1	21.7	22.4	16.0	20.6
350+°C	37.6	32.1	36.6	23.2	33.3	32.2	49.0	43.7
Cracker Feed (370–520°C):								
% (volume)	21.5	19.6	27.2	14.6	21.4	23.1	22.7	18.9
Sulphur % (weight)	0.71	0.21	0.28	0.29	0.13	0.39	1.94	2.19
K-factor	n.a.	11.9	11.8	12.1	12.3	12.2	11.8	11.9

Sources : Various oil companies, *Oil & Gas Journal*

(b) *Brent, Forties, Statfjord* and *Beryl* with Zueitina and Es Sider (Libya), Zarzaitine (Algeria), Bonny Medium and Forcados (Nigeria), WTI (USA).
(c) *Ninian* with Amna, Sarir and Bu Attifel (Libya).
(d) *Flotta* with Arabian Light (Saudi Arabia), Urals (USSR), Isthmus (Mexico), and Iranian Light.

Fulmar would fall between sets (a) and (b). Set (d) includes most of the light and fairly sweet crudes from the Gulf (Qatar, Abu Dhabi, Oman, Arabian Berri from Saudi Arabia and Kirkuk from Iraq). In our ordinal ranking this set is superior in quality to Flotta and inferior to Ninian.

This clustering defines the substitutability range of the North Sea crude blends with their main Atlantic Basin and Gulf competitors.

2.6   Conclusions

North Sea blends and crudes, with the exception of Ekofisk, display a fairly narrow range of physical and refining characteristics. This is partly a result of blending, a by-product of the pipeline transport system. They are similar to a wide variety of crudes normally traded in Europe; and some of the major blends, such as Brent and Forties, are very close indeed to important US crudes, mainly WTI. North Sea blends and crudes do not have any unusual properties of their own that cause difficulties in refining or confine them to specialist uses.

The market implications are important. As North Sea crudes satisfy the normal requirements of refiners, they are highly tradable and highly substitutable for other varieties. They do not raise particular demand problems of the type often reflected (in the case of certain Venezuelan crudes, for example) by the existence of very wide price differentials with other crudes (see Chapter 14).

Easy tradability throughout a fairly wide geographical area and similarity of properties have established an interface in the US market between Brent and Forties on the one hand and WTI on the other, and an even wider interface with Gulf and African crudes in Europe. These links, as we shall see later, contribute to the integration of the international petroleum market and to a greater convergence of oil price movements in different parts of the world.

CHAPTER 3

THE STRUCTURE OF THE NORTH SEA CRUDE OIL MARKET: PRODUCERS, SELLERS, BUYERS

3.1 Introduction

The purposes of this chapter are:

— to identify and characterize the economic agents operating in the North Sea crude oil market, namely producers, operators, sellers and end-use buyers;
— to assess the relative importance of their activities measured in terms of average annual volumes of output, sales or end-use purchases;
— to analyse the structure of the market, particularly the degrees of concentration on the selling side and on the buying side.

We shall use the following definitions to identify agents involved in the market:

— *Licensee*: a company holding a licence for part or the whole of either a producing or a non-producing UKCS/Norwegian field, or for blocks not yet explored.
— *Producing company*: a company holding a licence for part or the whole of a producing oilfield.
— *Operator*: a company responsible for the development and production of a field. The operator acts on behalf of the co-licensees.
— *Primary seller*: a company which is the first to dispose of oil produced in the North Sea through arm's length sales to third parties or through internal transfers. Primary sellers include all producing companies that do not assign their output entitlements at the production point to another company such as the operator or BNOC. (Some small companies find it convenient for logistic or cash-flow reasons to assign their output entitlement to the operator, BNOC or some other oil company. These producing companies are *not* primary sellers according to our definition.) BNOC and OPA are also primary sellers – though not producing companies – since they are the first sellers of oil

received, as it were, at source under participation and royalty arrangements.
— *Primary availabilities*: the volume of oil that a primary seller has at its disposal from its production entitlement (after royalty and participation) and from assignments. Volumes purchased from any source and for whatever purpose are not part of primary availabilities.

A schematic diagram showing the direction of transaction flows between the main groups of agents is given in Figure 3.1. The market may be visualized as an area with two borders. At one border, primary sellers deliver the initial supply of wet barrels extracted from the North Sea. At the other border end-users (refiners and holders of crude oil inventories, for example, the Strategic Petroleum Reserve) are the final recipients of crude oil volumes delivered to the market. In this chapter we are only concerned with the structure of the industry and the concentration of agents at these two borders of the market.

The market itself is taken to be the large area in between these borders. Not all primary availabilities are necessarily traded on the market since integrated companies have the option of transferring some of their oil internally. Furthermore, not all transactions flow directly from primary sellers to end-users. Primary sellers may sell oil to each other and to traders as well as to end-users. They may also purchase oil from the market for resale. In other words, there is a complex web of secondary sales and purchases of both wet barrels and paper contracts (claims on physical oil to be delivered at a future date which can themselves be sold/purchased from the moment they are initiated until the maturity date). The market for North Sea crudes involves term, immediate spot and forward transactions. It is a very active market in which a large number of agents are engaged. We shall focus on the functions, operations and structure of the complex market area that lies between primary sellers and end-users in Part III. This will concentrate on one segment of that market, commonly referred to as Brent, because trading in Brent has developed into a quasi-institutionalized set of spot and forward markets dealing with both wet and paper barrels. Brent plays a barometric role in the pricing of other North Sea crudes and may have an influence on economic behaviour in the world petroleum scene outside the North Sea.

Meanwhile, it is important to examine the wider market framework in which all North Sea crudes are traded; and to this task we now turn.

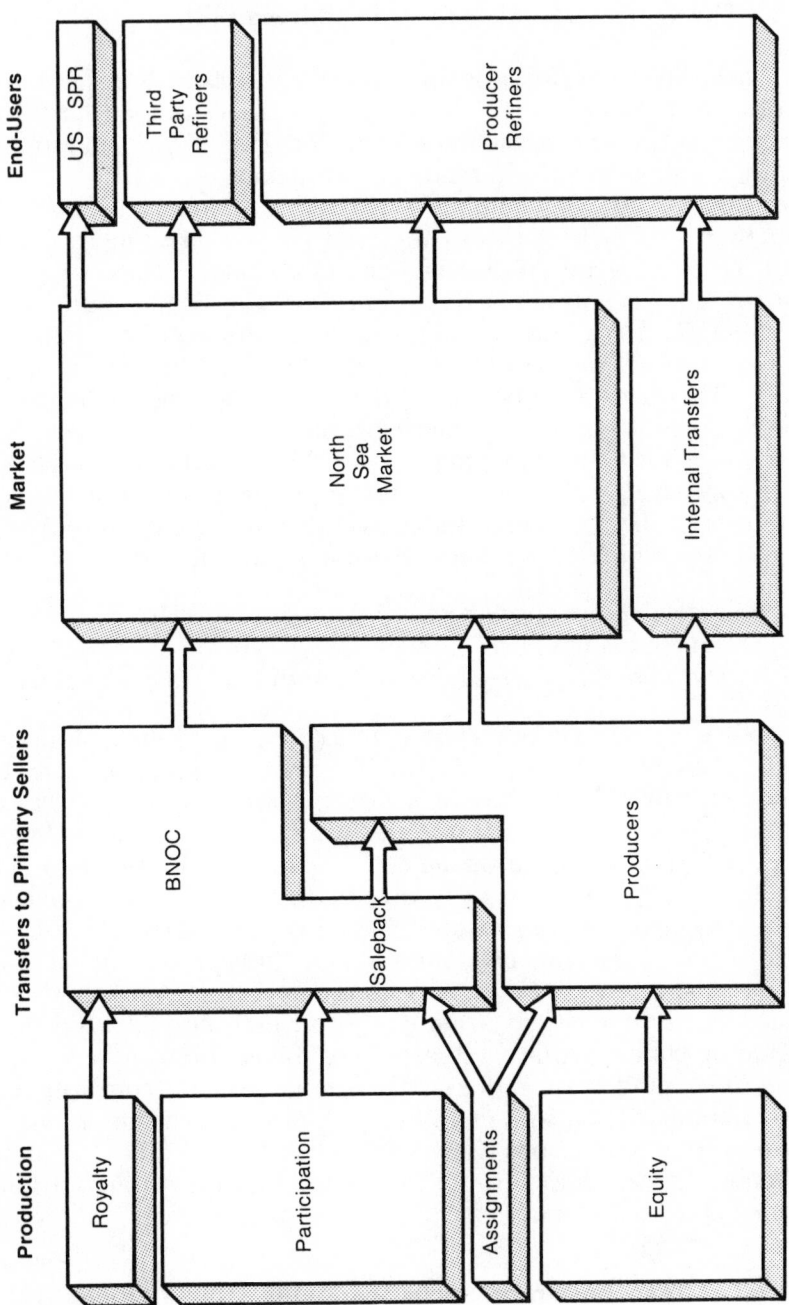

Figure 3.1   North Sea Market: Transaction Flows on the UKCS in 1984

3.2   The Concession Pattern of Producing Oilfields

The licensing of exploration and production in the UK has favoured widespread dissemination of concession shares among large and small oil corporations, oil and non-oil concerns, British and foreign entities. In Norway the spread is much narrower.

Complete lists of companies with licences in producing fields together with their shares of the field are presented in Tables S.7 and S.8 in Annex 8 (Statistical Appendix). Table S.7 is a complete list of the twenty-six UKCS and ten Norwegian oilfields in production at the end of 1984 (excepting Deveron whose average output spread over 1984 was a mere 2000 b/d because of a late start). The fields are listed in alphabetical order and the names of the licensees are given together with their respective shares.

Table S.8 lists the companies with holdings in the relevant fields.

It appears that the total number of companies with production interests in the thirty-six oilfields considered is eighty-one. (Moray Petroleum, which is also listed, has a net profit interest):

— Sixty-two companies have licences only in the UKCS.
— Seven companies have licences only in Norway.
— Twelve companies have licences in both the UKCS and Norway.

Producing companies include, among others, all the majors (BP, CFP, Chevron, Exxon, Gulf, Mobil, Shell and Texaco); many US independents (Amoco, Conoco, Getty, Marathon, Murphy, Occidental, etc.) with some notable exceptions (e.g. Atlantic Richfield, Sun, etc.); important European oil companies (Agip, Deminex, Elf-Aquitaine, Hispanoil, Norsk Hydro, Petrofina, etc.); the national oil corporation of Norway (Statoil); the privatised offshoots of two British national companies (Britoil and Enterprise Oil); oil companies that owe their emergence to the development of the North Sea (e.g. Lasmo, Clyde, etc.); and several non-oil companies including banks (Kleinwort Benson), insurance companies (Norwich Union), mining corporations (RTZ), large holding concerns (e.g. International Thomson Organisation), and important industrial corporations (e.g. Dow Chemical and ICI). An OPEC member country, Kuwait, has a direct producing interest through a fully-owned company (Santa Fe).

3.3   The Distribution of North Sea Fields Among Operators

The number of operators in the North Sea producing fields is much smaller than that of licensees or producers, and even smaller than the number of producing fields. All operators are oil companies of significant size and recognised experience. While we identified eighty-one entities with output shares in the producing fields of the UK and Norway, the number of operators was a mere thirteen.

This concentration of operations in a few hands does not have great significance for the structure of the market. The operator has a certain amount of influence on the development plan, on the investment timetable and probably on the production profile of the field. It should not be forgotten, however, that the operator acts on behalf of the co-licensees who have the right to be consulted and informed. The operator's autonomy depends within fairly narrow margins on the identity of the co-licensees. There may be more autonomy when the partners are non-oil companies or new entrants to the petroleum industry than when they are oil concerns with significant technical capabilities.

The list of operators for each field is given in Table S.9 in Annex 8. The distribution by volume of operation (measured in output terms, 1984) is shown in Table 3.1.

Table 3.1: UK and Norway. Operators of Producing Oilfields and Volumes of Operation. 1984. Thousand Barrels per Day. Percentages.

Operator	Volume of Operation			
	UK	Norway	Total	% Share
1. Shell	758	–	758	23.3
2. BP	546	–	546	16.8
3. Mobil	89	384	473	14.6
4. Phillips	77	260	337	10.4
5. Occidental	282	–	282	8.7
6. Chevron	237	–	237	7.3
7. Conoco	179	27	206	6.3
8. Britoil	135	–	135	4.2
9. Amoco	67	52	119	3.7
10. Marathon	88	–	88	2.7
11. Texaco	29	–	29	0.9
12. Union	24	–	24	0.7
13. Hamilton	18	–	18	0.5

The four largest operators are Shell, BP, Mobil and Phillips. Although Esso does not appear in the list, it should be recalled that it is Shell's fifty-fifty partner in all their UKCS fields. This ensures that the relationship between Esso and Shell as an operator is much closer than between co-licensees in other fields and their operators. Shell operates almost one quarter of North Sea oil output and 30 per cent of UKCS production. The four largest operators account for 65 per cent of the North Sea. In the UK, Occidental and Chevron are third and fourth respectively. With Shell and BP they account for 72 per cent of UKCS oil output. In Norway there are only four operators, but Mobil and Phillips have the lion's share.

3.4   The Distribution of Output among Producing Companies

Producing companies may be ranked by the volume of crude oil output (1984) attributable to their interests in the relevant oilfields. Table 3.2 presents the relevant data. Production is disaggregated between the UKCS and Norway and the total North Sea volume is also given.

The distribution of output per producing company is very skewed. A small number of companies accounts for the bulk of North Sea production. This can be shown as follows:

— The four largest producing companies – BP, Shell, Esso and Statoil – account for almost half the North Sea oil production as their combined share is 47.5 per cent.
— The eight companies with an output in excess of 100,000 b/d have a combined share of 63 per cent.
— The first ten companies account for 68.6 per cent of total production.

The top group of ten companies is comprised of four majors, three large US independents, one European oil company and two companies that owe their existence to the North Sea: the national oil corporation of Norway, Statoil, and the privatised offshoot of BNOC, Britoil.

Most of the non-oil companies with production interests in the North Sea produce very small volumes. A notable exception is the International Thomson Organisation which has a sizeable output of 56,000 b/d, larger than the production volume of certain majors such as Texaco and Chevron and of many important oil companies.

Table 3.2: UK and Norway. 1984. Oil Output by Producing Company. Thousand Barrels per Day

Company		Total Output	UK	Norway
1.	BP	494	494	–
2.	Shell	424	377	47
3.	Esso	418	377	41
4.	Statoil	206	–	206
5.	Britoil	160	160	–
6.	Mobil	120	58	62
7.	Phillips	118	26	92
8.	Conoco	104	63	41
9.	Petrofina	97	22	75
10.	Occidental	90	90	–
11.	Getty	64	64	–
12.	Gulf	63	63	–
13.	Allied	56	56	–
	Int Thomson Org	56	56	–
15.	Amerada Hess	54	34	20
16.	Agip	48	16	32
	Texas Eastern	48	28	20
18.	Texaco	47	47	–
19.	Deminex	46	46	–
20.	ICI	44	44	–
21.	Amoco	43	23	20
22.	Chevron	41	41	–
23.	Marathon	33	33	–
24.	Enterprise Oil	32	32	–
25.	Lasmo	27	27	–
26.	Elf	26	6	20
27.	Kerr McGee	19	19	–
	Norsk Hydro	19	2	17
29.	Murphy	17	17	–
	Odeco	17	17	–
31.	Santa Fe	15	15	–
32.	Ranger	14	14	–
33.	BC Resources	12	12	–
	Ultramar	12	12	–
35.	CFP	10	–	10
	Charterhouse	10	10	–
	Century Power Light	10	10	–
	Tricentrol	10	10	–

Table 3.2:   Continued

The following companies have less than 10,000 b/d each:

Louisiana Land, NOCO, RTZ	9
Saga, Sovereign, Tenneco, Union	8
Burmah	7
Hamilton, Trafalgar House	6
Dow Chemical, Hunt	5
Bow Valley, OK Exploration	4
Clyde, Sulpetro, Transworld	3
Associated News, Charterhall, DNO	
North Sea and General	2
AB Exploration, Anvil, Berkeley Exploration,	
Coalite, Cofranord, Coparex, Eurafrep,	
Floyd, Goal Petroleum, Hispanoil,	
Ind Scot Energy, Kleinwort Benson,	
Lochiel, (Moray Petroleum), Norwich Union,	
Pict, Plascom, Reading and Bates, Saxon,	
Texas Gas, Third Triton, Union Jack,	
Viva	1 or less

Source: Wood Mackenzie

3.5   The Distribution of Output Among Primary Sellers

Producing companies do not have their entire output available to them for free disposal outside the boundaries of the North Sea. In the UK, before the abolition of BNOC, the amounts available to a producing company could be expressed as follows:

Availability   =   Production *minus* Royalties in kind *minus* Participation Sales to BNOC *plus* Buy-backs *plus or minus* Assignments to or from other North Sea producing companies.

As mentioned before, primary sellers include all producing companies which do not assign their output entitlements to another company as well as BNOC/OPA. It must be stressed that in this context 'availability' refers only to the volume of oil that a producing company has at its disposal before it begins lifting or trading. It is essentially its output net of the amounts paid or sold to the state and adjusted for quantities acquired from or assigned to other producers in the fields.

For these reasons 'buy-back' is defined here in the narrow and strict sense of buy-back of participation crude. Some companies were able to repurchase from BNOC amounts in excess of their participation crude. These additional amounts were contract sales rather than strict buy-backs. They could be phased out or increased as in any term contract. The true buy-back of participation crude was conceived on the principle of 'same barrels' so that no physical deliveries were ever involved. The repurchase of royalty oil was not based on that principle. The barrel received need not be identical to the barrel supplied for royalty payment. Physical deliveries could be involved.

Similarly, assignments are defined very narrowly; swaps, exchanges and sales taking place outside the producing fields are ordinary transactions, not assignments.

BNOC/OPA is the only non-producing company with primary availabilities, which consist of royalties in kind, assignments and participation sales minus sales-back.

The estimates of availabilities of UKCS crudes to primary sellers presented in Table 3.3 are derived on the basis of the following assumptions:

— Total production by company in 1984 is as shown in Table 3.2.
— Royalties are deducted from production and attributed to BNOC. BNOC received royalties in kind from all producing fields except Argyll, Auk, Beatrice, Buchan, Heather, Maureen, Montrose and Tartan. Licensees in these eight fields retained royalty oil and paid their dues in cash. The notional royalty rate is 12.5 per cent but allowances are set against royalties (see Chapter 8) and actual rates vary from field to field. The actual average royalty rate in 1984 was approximately 10 per cent and actual deliveries of royalties in kind were made at an average rate of 11.3 per cent. The Department of Energy reimbursed the difference (about £250 million) in cash to the relevant companies.
— Participation crude minus buy-back is part of BNOC's availabilities.
— The participation rate is 51 per cent of production net of royalties except in the following cases:

(a) Beryl: 45.5 per cent for Mobil
(b) Brae: 38.75 per cent for all licensees except Britoil
(c) Dunlin: 26.50 per cent for Conoco and Gulf
(d) Fulmar: 34 per cent for Amoco and Mobil

(e) Hutton: zero rate for Conoco and Gulf and 34 per cent for Mobil and Amoco
(f) North West Hutton: as for Fulmar
(g) Montrose: 29.2 per cent for Amoco
(h) Murchison (UK): 16 per cent for Conoco and Gulf
(i) Statfjord (UK): 26.5 per cent for Conoco and Gulf

— Companies with buy-back agreements in 1984 (in the narrow sense of the concept) were Amoco, BP, Esso, Mobil, Shell and Texaco. Participation/buy-back agreements tended to be different from company to company; they appear to have been unnecessarily complex and shrouded with secrecy. Nevertheless it can be assumed that in 1984 all the companies mentioned retained 100 per cent of their participation oil.
— Other repurchase agreements with BNOC are treated as long-term sales contracts. This means that the volumes involved are part of BNOC's availabilities, not part of the companies'.
— Most small licensees assigned their oil to BNOC or to the operator of the field or to some other oil company. This means that they did not take delivery of their share of the fields' outputs, and did not themselves bring oil onto the market. The amounts assigned are deducted from their availabilities and added to that of the assignee in each case.
— Assignments to BNOC (1984) are thought to include:

(a) All the production entitlements of Enterprise Oil
(b) Burmah's and Charterhouse's entitlements from Thistle
(c) Hunt's from Beatrice
(d) Those of AB Exploration, Agip, Anvil, Coalite, Dow, Floyd, North Sea and General, Sovereign, Texas Gas, Third Triton and Transworld from Claymore
(e) Those of Berkeley Exploration, Charterhall, Charterhouse, CPL, Hispanoil, Ind Scottish Energy, North Sea and General, OK Exploration, Plascom, Saxon, Sovereign, Union Jack and Viva from Forties

— Assignments to BP (1984) are thought to include:

(a) The entitlements of Charterhall, Clyde, Goal and Tricentrol from Buchan
(b) Those of Norwich Union and Trafalgar House from Forties

— Other assignments include Ranger (Ninian) to Chevron; International Thomson (Piper) to ICI; Sovereign and Norsk Hydro (Brae) to Marathon; and Bow Valley (Brae) to BC Resources and Louisiana Land.

Table 3.3:   UK and Norway. Availabilities of North Sea Oil by Company (Primary Sellers). 1984. Availabilities after BNOC's Abolition at 1984 Production Levels. Thousand Barrels per Day.

Company	Net Availabilities				
	1984 UK	*1984 Norway*	*1984 Total*	*Without BNOC*	*Difference*
BNOC	805	–	805	–	(805)
OPA	–	–	–	287	287
BP	439	–	439	445	6
Shell	330	40	370	370	–
Esso	330	35	365	365	–
Statoil	–	268.5	268.5	268.5	–
Mobil	51	52	103	103	–
Phillips	13	83	96	109	13
Conoco	44	35	79	90	11
Petrofina	11	67	78	90	12
Britoil	69	–	69	141.5	72.5
Texaco	46	–	46	46	–
Gulf	44	–	44	55	11
Amoco	20.5	18	38.5	38.5	–
Occidental	38.5	–	38.5	78.5	40
Agip	6.5	29	35.5	44.5	9
ICI	34.5	–	34.5	70	35.5
Amerada Hess	15	18	33	48	15
Texas Eastern	12	18	30	42	12
Getty	27.5	–	27.5	56	28.5
Allied	24	–	24	49	25
Chevron	23	–	23	47	24
Elf	3	18	21	23.5	2.5
Marathon	21	–	21	34	13
Deminex	20.5	–	20.5	42	21.5
Norsk Hydro	–	15	15	15	–
Murphy/Odeco	15	–	15	30	15
Lasmo	12	–	12	25	13
CFP	–	9	9	9	–
Kerr McGee	9	–	9	18	9
Int Thomson Org	9	–	9	17.5	8.5
Noco	–	8	8	8	–
BC Resources	6	–	6	10	4
Saga	–	6	6	6	–
Santa Fe	6	–	6	13	7
Ultramar	5.5	–	5.5	11	5.5

The availability of North Sea crudes to producing companies in Norway is much easier to estimate.

Availability = Production minus Royalty.

The royalty rates were 10 per cent for Ekofisk and Valhall, 8 per cent for Murchison and 16 per cent for Statfjord. The very small licensees – Cofranord, Coparex and Eurafrep – are thought to have assigned their production to Elf.

The companies listed in Table 3.3 are the main primary sellers of North Sea crudes. (Very small primary sellers are not listed.) They have primary access to these crudes, and can be considered as the original sources of supplies. This group of companies constitutes the supply side of the North Sea market. It can be seen that primary sellers are less numerous than companies with a claim on output. Furthermore, the distribution of the total amount of oil available (equal to total production) among these companies is significantly different, at least in the UK, from the distribution of output. The reason is that the output distribution is reshuffled between producing companies and BNOC through the complex royalty, participation and buy-back arrangements. Finally, assignments reduce the number of players from about seventy producers to a maximum of fifty primary sellers.

The number of primary sellers and the distribution of total availabilities among them determine the structure of the market on the selling side. In 1984, and until the abolition of BNOC, this distribution appears to have been very unequal. The relevant data for 1984 are presented in Table 3.3.

Consider the supply structure in 1984. In the North Sea as a whole, five companies dominated the market. In order of importance these were:

— BNOC, BP, Shell, Esso and Statoil.

These five concerns had at their disposal almost 70 per cent of a total North Sea production of 3.25 mb/d. Their respective shares were as follows (individually and cumulatively):

BNOC	24.8%	24.8%
BP	13.5%	38.3%
Shell	11.4%	49.7%
Esso	11.2%	60.9%
Statoil	8.3%	69.2%

More striking is the fact that the first three (BNOC, BP and Shell) taken together accounted for approximately half the supply side of the market.

The five largest primary sellers of UKCS crudes were:

— BNOC, BP, Shell, Esso and Britoil.

Their respective shares were as follows (individually and cumulatively):

BNOC	31.8%	31.8%
BP	17.4%	49.2%
Shell	13.0%	62.2%
Esso	13.0%	75.2%
Britoil	2.7%	77.9%

The concentration was greater for the UKCS than for the North Sea as a whole. Almost half the production was in the hands of two companies only – BNOC and BP. The first four accounted for three-quarters of the market.

In Norway, the five largest primary sellers were:

— Statoil, Phillips, Petrofina, Mobil and Shell.

Their respective shares were as follows (individually and cumulatively):

Statoil	37.2%	37.2%
Phillips	11.5%	48.7%
Petrofina	9.3%	58.0%
Mobil	7.2%	65.2%
Shell	5.5%	70.7%

The structure in Norway is in some respects different from that of the UK. The top company in Norway (Statoil) commands a larger share than the top one in the UK (BNOC); but the cumulative shares from the second to the fifth company are smaller in Norway than in the UK.

A better measure of concentration is given by the Herfindahl index which takes into account the number of companies in the market and their shares. The index is:

$$H = \Sigma s_i^2 , 0 < s_i < 1$$

where s_i is the share of company i. The H-index has a simple and attractive interpretation. It can be translated into the number of equal-sized companies that would yield the same degree of market concentration as the actual distribution of companies with unequal shares. This number is the inverse of H.

The degree of concentration is a preliminary indication of the extent of competition (or the lack of it) in a market. A small equivalent number of equal-sized firms (1/H) suggests that a monopoly or an oligopoly may exist. A high number suggests that the market is competitive.

The Herfindahl indices for the supply side of the UK, Norwegian and total North Sea crude oil markets are as follows:

	H-index	Equivalent number of equal-sized firms (1/H)
UK	0.17655	5.66
Norway	0.17789	5.62
Total North Sea	0.12155	8.23

These results can be taken to mean that:

— The degree of concentration on the supply side of the market is virtually the same in the UKCS and in Norway.
— This concentration is fairly high. It is equivalent to that of a market in which supplies are provided in equal shares by five or six firms.
— The degree of concentration is lower when the UKCS and Norway are considered as a single industry. The equivalent number of equal-sized firms rises to just over eight. This is because the actual shares of firms in the aggregate market are smaller, and the number of firms is higher than in either the UKCS or Norway taken separately.

3.6   Primary Availabilities of the Main North Sea Blends

The same methods of calculation have been used to estimate the net availabilities per company of the main North Sea blends of crude oil, namely Brent, Forties, Ninian and Flotta in the UK, and Ekofisk in Norway.

Table 3.4 contains the information for 1984, and for a world without BNOC in which total UKCS oil output is assumed to be equal to 1984 production.

In 1984, the bulk of Brent blend, more precisely 82 per cent, was supplied by only three companies (Esso, Shell and BNOC). Forties, and to a lesser extent Ninian, supplies were dominated by two companies (BP and BNOC). BNOC had the lion's share of Flotta supplies (53.8 per cent) with the rest almost equally divided among four or five oil companies. BNOC had the largest availabilities of both Ninian and Flotta blends but did not occupy such a prominent position in its access to Brent and Forties. This different pattern of availabilities was due to the participation/buy-back arrangements which enabled companies such as BP, Esso and Shell (the main licensees of Brent and Forties) to retain their participation oil, but did not give buy-back rights to the smaller companies, which dominated the fields in the Flotta system.

The concentration index measured as the equivalent number of equal-sized firms (inverse of Herfindahl) gives the following results:

Field	1/H
Ekofisk	6.2
Brent	4.4
Ninian	3.6
Flotta	2.9
Forties	2.2

It is interesting to note the contrast between the structure of availabilities of Ekofisk and the UKCS blends. The degree of concentration is lower in the case of Ekofisk than in all others.

Though the number of fields involved in some of the UKCS blend systems and the number of licensees are both fairly large, the supply of these blends is concentrated to a high degree. The apparent diversity of suppliers is reduced by the index to between two and five firms.

3.7 The Abolition of BNOC and the Market Structure

The abolition of BNOC will change the distribution of availabilities among primary sellers very significantly, and therefore the structure and concentration of the market on the supply side.

The most obvious change is the removal of the largest supplier with some 800,000 b/d at its disposal, and its replacement by a much smaller entity. The less obvious, but more interesting, point to note is that the abolition of BNOC does not add a single barrel to

Table 3.4:   Net Availabilities of North Sea Blends by Main Company (Primary Sellers). 1984. Availabilities after BNOC's Abolition at 1984 Production Levels. Thousand Barrels per Day.

Company	1984		Without BNOC	
	Volume	% share	Volume	% share
1. Brent				
Esso	265.4	29.8	265.4	29.8
Shell	265.4	29.8	265.4	29.8
BNOC/OPA	199.8	22.4	108.0	12.1
Conoco	30.8	3.5	36.1	4.0
Gulf	28.3	3.2	33.6	3.8
Britoil	23.4	2.6	47.6	5.3
Others	(78.2)	(8.7)	(135.2)	(15.2)
Total	891.3	100.0	891.3	100.0
2. Forties				
BP	305.5	60.7	308.7	61.3
BNOC/OPA	123.3	24.5	63.0	12.5
Marathon	20.8	4.1	33.9	6.7
Others	(54.1)	(10.7)	(98.1)	(19.5)
Total	503.7	100.0	503.7	100.0
3. Ninian				
BNOC/OPA	142.1	38.4	43.2	11.7
BP	125.2	33.8	125.2	33.8
Chevron	23.1	6.2	47.3	12.8
Britoil	21.7	5.9	44.3	12.0
ICI	18.8	5.1	38.4	10.4
Others	(39.0)	(10.6)	(71.7)	(19.3)
Total	369.9	100.0	369.9	100.0
4. Flotta				
BNOC/OPA	167.3	53.8	35.3	11.3
Occidental	38.5	12.4	78.6	25.3
Texaco	29.0	9.3	29.0	9.3
Getty	27.4	8.8	56.3	18.1
Allied	24.2	7.8	49.3	15.9
Others	(24.4)	(7.9)	(62.3)	(20.1)
Total	310.8	100.0	310.8	100.0

Table 3.4:   Continued

Company	1984	
	Volume	% share
5. Ekofisk		
Phillips	83.1	26.7
Petrofina	67.4	21.6
Statoil	31.4	10.1
Agip	29.2	9.4
Elf	18.0	5.8
Norsk Hydro	14.9	4.8
Amerada	14.3	4.6
Amoco	14.3	4.6
Texas Eastern	14.3	4.6
Others	(24.9)	(8.0)
Total	311.8	100.0

the availabilities of the big oil companies, BP, Shell, Esso and Mobil, which have the largest shares of North Sea production. The reason is that these companies had buy-back arrangements and already retained their participation oil.

BNOC oil will go to the other companies. Those with very small availabilities will not gain much in absolute weight (though the amounts of oil at their disposal will double in many cases) because the number of barrels involved is not very significant. The medium-sized companies will acquire most of the BNOC participation oil and will gain the most weight. Their place and significance in the structure will alter considerably, and these transformations will lead to changes in the degree of concentration. They are also likely to lead to changes in conduct and behaviour of the companies concerned, but these are more difficult to ascertain in advance.

The changes in structure after the abolition of BNOC can be assessed in a fairly easy way. The method adopted takes the 1984 output as the basis of calculations made for the purpose of illustrating changes. Participation crude is reallocated to the original producing companies and royalties are transferred to OPA. The amounts assigned to BNOC by small companies are assumed to be retained by these companies after the abolition. Other assignments (e.g. to BP, ICI, etc.) continue as before. The resulting net availabilities, under these assumptions (Alternative I), are shown in Table 3.3.

The amount redistributed is BNOC's net availability of 805,000 b/d minus royalties of 287,000 b/d, that is 518,000 b/d. The main beneficiaries of the reallocation, in order of importance, are:

Company	Volume ('000 b/d)	%
Britoil	72.5	14
Occidental	40.0	8
ICI	35.5	7
Getty	28.5	5.5
Enterprise	28.0	5.5
Allied	25.0	5
Chevron	24.0	5
Deminex	22.0	4
Murphy	16.0	3
Amerada Hess	15.0	3
Lasmo	13.0	2.5
Marathon	13.0	2.5
Phillips	13.0	2.5
Others	(172.5)	(33.0)
Total	518.0	100.0

Some commentators have suggested that the abolition of BNOC will reduce the significance of the North Sea spot market. The large volumes of oil which BNOC used to sell on spot in late 1984/early 1985 will be returned to oil companies which may retain it for their own use. This argument overlooks the following facts:

— The abolition does not affect the large integrated companies with refineries in the UK such as Amoco, BP, Esso, Mobil, Shell and Texaco, because these companies already retain all their participation crude under buy-back arrangements.
— The companies that will obtain most of BNOC's oil are oil companies, some of which are entirely without refineries (e.g. Britoil which is also the largest beneficiary) and many of which are without refineries in the UK or North West Europe (Occidental, Allied, Marathon, etc.).
— A large share of BNOC's oil (some 35 per cent) will remain with OPA and the remainder (between 20 and 25 per cent) will be returned to very small oil companies and to non-oil concerns. Under present market conditions, all these volumes are likely to be disposed of on the spot market.

A major effect of BNOC's abolition is on the structure of the North Sea market. The abolition significantly reduces the degree of concentration. We have calculated Herfindahl indices of concentration (the supply side of the market) for the North Sea after the abolition of BNOC. Given the assumptions of Alternative I we find that the degree of concentration declines considerably in the UK and consequently in the North Sea as a whole. The degree of concentration measured by the equivalent number of equal-sized firms changes as follows:

	Before the abolition	After the abolition
UK	5.66	10.94
Total North Sea	8.23	14.26

Before the abolition of BNOC the market structure (UK) resembles an oligopoly of five or six equal-sized firms. After the abolition the notional structure involves eleven equal-sized firms.

Alternative assumptions about the redistribution of oil supplies after the abolition of BNOC can be made. For example, we may realistically assume that the volumes previously assigned to BNOC by small companies would not be retained by them but assigned to the operators of the relevant fields. We refer to these assumptions as Alternative II. In such an instance some 50,000 b/d might be reallocated, mainly to BP (some 20,000 b/d) and to Occidental (about 15,000 b/d). Britoil, Marathon, and Hamilton would also obtain small additional volumes.

The abolition of BNOC has interesting implications for companies such as the International Thomson Organisation (ITO) and Lasmo, to take two notable examples. ITO is known to assign its entitlement from Piper to ICI, and to sell its entitlement from Claymore to Neste Oy. In 1984 ITO produced 56,000 b/d and was left with 25,000 b/d after royalty and participation. After the abolition of BNOC, ITO would be entitled to 49,000 b/d. This is a considerable amount of oil and might justify the setting up of an ITO trading operation. The interesting question is whether ITO would maintain present arrangements (this would double ICI and Neste's purchases volume and probably exceed their requirements) or decide to market this substantial volume directly and diversify its outlets. The same question may be asked of Lasmo whose availabilities would increase from 12,000 to 25,000 b/d.

Another realistic possibility, recently much discussed in the Press, is the emergence of an intermediary organization to pool the small amounts available to smaller companies and sell the sizeable volume that results on their behalf. The intermediary could be an existing oil company with trading experience in the North Sea, a trader, or a new private agency set up specifically for the purpose.

We found that the concentration index is not very sensitive to these changes in assumptions. The reassignment of, say, 50,000 b/d to BP, Britoil, Occidental, etc. (Alternative II) does not significantly change the value of the concentration index as measured under the assumptions of Alternative I.

To sum up; the abolition of BNOC will undoubtedly have an impact on the supply structure of the North Sea market. Medium-sized companies, some of which are without downstream interests (Britoil, ITO, Lasmo, etc.), are the main beneficiaries of the reallocation of BNOC's oil. Furthermore, the new entity that replaces BNOC will still have access to a large volume of oil. OPA will rank fourth in terms of net availabilities, just after BP, Shell and Esso with some 280–290,000 b/d of crude oil at its disposal. Yet, OPA and the non-integrated recipients of BNOC oil will continue to supply their availabilities to the third-party market since they have no internal demand to satisfy. The abolition of BNOC is thus unlikely to alter significantly the current allocation of North Sea crudes among spot sales, third-party term contracts and internal disposals.

We have also seen that the degree of market concentration as measured by the H-index will decline. However, this result should be interpreted cautiously. The apparently high degree of concentration that obtained before the abolition was partly due to the very high share of BNOC in total availabilities. Yet BNOC did not always wield market power to the extent suggested by its share of total supplies. For that reason, the change in concentration as measured by the H-index before and after the abolition does not necessarily indicate a similar change in the degree of market power on the supply side of the UKCS market.

3.8   The Buyers (End-users) of North Sea Crudes

A detailed pattern of North Sea crude purchases by company has been established for 1984. This enables us to describe the structure of the market on the buying (end-user) side fairly accurately, and to compare it with the features of the selling side which have already been described.

The method adopted to draw the buying picture was painstaking and long but it has the advantage of yielding reasonably accurate results. The data used are:

— US oil import statistics by company compiled monthly by the American Petroleum Institute
— Shipping data from *Lloyd's List* and the *Shetland Times*
— Lifting schedules from a variety of British and Norwegian industry sources
— Refinery data on ownership, capacity, rate of utilisation and throughput of refineries in North West Europe from both published and unpublished sources

The method followed to identify buyers of North Sea crudes and the approximate volume of their purchases is similar to a tanker tracking exercise. We use, however, a number of cross-checks (production, net availabilities, refinery throughput, etc.) to ensure that the pattern derived from the shipping data is broadly correct. Our data on liftings from onshore terminals in the UK (i.e. data for liftings of Brent, Ninian, Forties and Flotta) are close to the true figures since they are derived directly from accurate sources.

Liftings from offshore terminals, which represent a small proportion of UKCS output, are estimated indirectly and are therefore less reliable. The liftings shown for each company therefore include two components: an estimated and a known amount. It follows that the margin of error is smaller for companies that lift a large proportion of their crude from onshore terminals. For this reason we feel happier about our figures for the UKCS liftings of Esso, Shell and Texaco, who acquire a very high proportion of their requirements from onshore terminals, than those for BP, Mobil and Conoco, for whom this proportion is smaller. We also believe that our data on UKCS liftings for CFP, Petrofina, Elf, KPC, Gulf, ICI, Veba, Norsk Hydro and Neste Oy are fairly accurate because these companies acquired all their UKCS oil from onshore terminals.

Table 3.5 presents detailed information on North Sea oil purchases by company for 1984. It lists thirty-one oil companies which lifted an average of 10,000 b/d or more in that year and ranks them in order of importance. The table distinguishes between liftings to European and US destinations; it also disaggregates the volume purchased between the UKCS and Norway, and between UKCS onshore terminals and offshore fields (the latter includes Beatrice, however, which loads on shore at Nigg Bay).

Table 3.5:   UK and Norway. Main Buyers of North Sea Crudes and Volumes Purchased. Thousand Barrels per Day. 1984

Company	Purchases for European Destinations				US Purchases			Grand Total
	UKCS on-shore	UKCS off-shore	Nor-way	Total	UKCS	Nor-way	Total	
Esso	476	43	110	629	3	–	3	632
Shell	417	72	106	595	–	–	–	595
BP/Sohio	200	90	55	345	34	52	86	431
Texaco	127	34	45	206	3	–	3	209
Mobil	11	51	66	128	14	1	15	143
CFP	98	–	17	115	–	–	–	115
Sun	–	–	–	–	108	7	115	115
Conoco	4	61	20	85	–	–	–	85
SPR	–	–	–	–	67	11	78	78
Elf	53	–	12	65	–	–	–	65
Amoco	31	6	–	37	17	8	25	62
Gulf	21	–	10	31	27	4	31	62
Statoil	–	–	50	50	–	–	–	50
Svenska OK	27	–	15	42	–	–	–	42
Veba	32	–	10	42	–	–	–	42
Wintershall	7	15	10	32	–	–	–	32
Petrofina	21	–	10	31	–	–	–	31
Svenska Pet	4	10	15	29	–	–	–	29
Phillips	2	–	15	17	–	11	11	28
KPC	17	–	10	27	–	–	–	7
Neste Oy	15	–	10	25	–	–	–	25
Champlin	–	–	–	–	17	8	25	25
INPC	22	–	–	22	–	–	–	22
URBK	11	–	10	21	–	–	–	21
Norsk Hydro	17	–	–	17	–	–	–	17
CEPSA	17	–	–	17	–	–	–	17
Murphy	8	–	–	8	8	–	8	16
Kerr McGee	–	–	–	–	14	–	14	14
ICI	9	–	5	14	–	–	–	14
Crown Diamond	–	–	–	–	4	7	11	11
Shamrock	–	–	–	–	10	–	10	10

We have also identified eighteen other companies that lifted less than 10,000 b/d on average in 1984 and these are listed in Table 3.6.

The coverage of these liftings data can be assessed as follows. The total volume of liftings identified for 1984 is 3.169 mb/d while

Table 3.6:   UK and Norway. Minor Buyers of North Sea Crudes and Volume
Purchased. Thousand Barrels per Day. 1984

Company	Total Amount Purchased	Company	Total Amount Purchased
Arco	8	Petroliber	3
Union	8	Citgo	3
Gatoil	7	La Gloria	2
Ashland	7	Petronor	2
Petrola	6	EMP	2
Mapco	6	Scanoil	2
Petrogal	6	Coastal	1
Farmer's Union	5	Koch	1
Getty	4	Texas City Refinery	1

actual production in 1984 was 3.250 mb/d. This very small
discrepancy can easily be explained by minor inventory changes and
overlaps of consecutive years at the beginning and at the end of
1984. The breakdown of total liftings is as follows (thousand barrels
per day):

UKCS:	Liftings to European destinations	2057
	Liftings to the USA	(358)
	Recorded imports to the USA	362
	Liftings to Canada	16
	Liftings to the Caribbean	11
	Total UKCS	2446
Norway:	Estimated liftings to European destinations	601
	Recorded imports to the USA	122
	Total Norway	723
	Grand Total	3169

The analysis of data presented in Tables 3.5 and 3.6 yields the
following conclusions:

— The top six buyers of North Sea crudes for European destinations (1984) are Esso, Shell, BP, Texaco, Mobil and CFP. They all belong to the old family of the Seven (actually, eight) Sisters, another unmistakable sign of the survival of their former dominance. The six account for 75 per cent of North Sea crude liftings to Europe. The concentration index gives an equivalent number of equal-sized firms of 7.3.

— The three major importers of North Sea crudes into the USA (1984) are Sun, the Strategic Petroleum Reserve and BP/Sohio. They account for 58 per cent of North Sea crude imports to the USA. Other major buyers are Gulf, Champlin and Amoco. The top six account for 75 per cent of these imports, and this share happens to be identical to that of the top six companies lifting North Sea crudes to European destinations. The concentration index for US importers gives an equivalent number of equal-sized firms of 7.8.

— The top six buyers of North Sea crudes for *all* destinations are the same six as for Europe. These are Esso, Shell, BP/Sohio, Texaco, Mobil and CFP. However, Sun should be added to this list and ranked in the sixth place together with CFP because the volumes lifted by these two companies happened to be exactly the same in 1984 (115,000 b/d). The ranking of the first six remains unaltered, but the six top companies account for 64 instead of 75 per cent of liftings. The concentration index gives an equivalent number of equal-sized firms of 9.2. Thus the concentration on the buying side of the North Sea market is slightly lower than the degree of concentration on the selling side before the abolition of BNOC (8.2 equal-sized firms). After the abolition, the buying pattern is expected to remain unchanged while the selling side should become less concentrated. There should thus be a reversal of the current situation leading to a relatively greater concentration of market power in buyers' than in sellers' hands.

3.9 North Sea Crudes Liftings and Refinery Throughput

The purpose of this section is to identify companies that are running proportionately more (or less) North Sea crudes in their refineries than the average North West European refiner in 1984. The method used compares each company's liftings of North Sea crudes with an estimated 'notional' throughput of North Sea crudes based on, first, the throughput required to load fully each company's upgrading capacity in each country, and, secondly, the proportion of North Sea crudes in each country's total imports. The exercise was performed for each company and country and the results for all countries were aggregated for each company.

Table 3.7 compares for each company its liftings of North Sea crudes to European destinations with its 'notional' throughput of North Sea crudes. The 'notional' throughput is that level that would be consistent with the average share of North Sea crudes in the oil imports of the relevant European countries.

Table 3.7: Crude Liftings of North Sea Crudes by Company Compared with Notional Throughput. North West Europe. 1984. Thousand Barrels per Day

Company	Liftings	Notional Throughput
(a) Above Notional		
Esso	629	470
Shell	595	517
BP	345	310
(b) Below Notional		
Mobil	128	174
CFP	115	150
Wintershall	32	45
Petrofina	31	76
Gulf	31	43
Belgian Ref Corp	–	15
Coastal	–	11

It appears that sixteen refiners lifted volumes of North Sea crudes approximately equal to the 'notional' or average through-put. These are:

— Amoco, Conoco, Elf, ICI, KPC, Murphy, Neste Oy, Norsk Hydro, Phillips, Statoil, Svenska OK, Svenska Petroleum, Texaco, URBK and Veba.

Companies lifting more than the 'notional' throughput level are BP, Esso and Shell; and companies lifting less than the 'notional' throughput are the Belgian Refining Corporation, Coastal, CFP, Gulf, Mobil, Petrofina and Wintershall.

Some companies seem to have a discernible preference for

certain North Sea crudes while others refine a fairly even mix of crudes. Companies which appear to have definite preferences are:

— BP: Forties (which is run at its Grangemouth refinery)
— Esso: Brent, Forties and Flotta
— Shell: Brent and Ninian
— Texaco: Forties
— CFP: Brent and Flotta
— Petrofina: Flotta
— Conoco: Maureen
— Elf: Ninian
— KPC: Forties
— Wintershall: Beatrice
— Mobil: Beryl and Norwegian crudes, probably Statfjord

Preferences of US importers for certain blends are revealed by the following data for 1984 (thousand barrels per day):

Type	US Imports	% of Total
Brent	152	42.0
Forties	78	21.5
Ninian	115	31.8
Flotta	–	–
Total	345	95.3
Total US imports of UKCS crudes	362	100.0

3.10   Sellers and Buyers (End-users) of Brent and Ninian Blends

This section examines the trading relationships between the major primary sellers of Brent and Ninian blends f.o.b. Sullom Voe during 1984. It draws on the results of our tanker tracking work and allows us to identify, with a reasonable degree of accuracy, both the volume of internal transfers within integrated companies (BP, Esso and Shell) and sales to specific third parties.

The main results are summarized in Tables 3.8 and 3.9. Table 3.8 shows the *quantities* sold by the four largest primary sellers – BNOC,

Table 3.8: Sales and Purchases of Brent and Ninian Blends at Sullom Voe. 1984. Thousand Barrels per Day

From: To:	BP (Ninian)	Esso (Brent)	Shell (Brent)	BNOC (Both)	Others (Both)	Total
BP	13	–	15	12	36	76
Esso	–	116	52	40	37	245
Shell	22	42	72	84	84	304
Others,Europe	44	32	37	96	141	350
Others,North America	43	43	57	96	53	292
Total	122	233	233	328	351	1267

BP, Esso and Shell – of Brent and Ninian at Sullom Voe, and the quantities purchased by four buyers – BP, Esso, Shell and 'Others' – in Europe and North America.

Table 3.9 shows the *percentages* of Sullom Voe oil sold by the same entities. Both tables refer to averages for 1984 arrived at by aggregating monthly data provided by the tanker tracking exercise. The tables reveal interesting features of inter-company trade which are described below.

Table 3.9: Distribution of Companies' Sales of Brent and Ninian Blends at Sullom Voe. 1984. Percentages

From: To:	BP (Ninian)	Esso (Brent)	Shell (Brent)	BNOC (Both)	Others (Both)
BP	10.6	–	6.4	3.7	10.2
Esso	–	49.8	22.3	12.2	10.5
Shell	18.0	18.0	30.9	25.6	24.0
Others,Europe	36.1	13.7	15.9	29.3	40.2
Others,North America	35.2	18.5	24.5	29.3	15.1
Total	100.0	100.0	100.0	100.0	100.0

In 1984 the three major integrated companies (BP, Esso and Shell) accounted for 588,000 b/d (46.4 per cent) of primary availabilities at Sullom Voe; and they lifted 625,000 b/d (49.4 per cent) of the total volume available. The overall imbalance between primary availabilities and liftings is not very large. However, these companies had much wider recourse to trading than is suggested by the small difference between their global requirements and their primary availabilities.

The first reason is that the small overall imbalance conceals larger imbalances for individual companies which partly cancel out when availabilities and requirements are aggregated. In 1984, BP appeared to be long on Sullom Voe blends (122,000 b/d availabilities against 76,000 b/d requirements); Esso was almost balanced (233,000 against 245,000 b/d respectively); and Shell was short (233,000 against 304,000 b/d).

The second reason is that integrated companies do not always retain their oil for their own use. They sell their oil in arm's length transactions and purchase part of their requirements from the market. An important motivation, as we shall see later, is tax spinning. The Sullom Voe trading matrices (Tables 3.8 and 3.9) give a clear indication of the extent of tax spinning. Companies' behaviour seems to vary in this respect. The most interesting part of the matrix consists of the four cells which describe the Esso/Shell self- and cross-trading relationships. Table 3.9 shows that in 1984 Esso retained half its Brent availabilities while Shell only retained 31 per cent. Furthermore, Esso sold 18 per cent of its Brent availabilities to Shell, and Shell sold 22.3 per cent of its own (and identical) Brent availabilities to Esso. These are genuine and perfectly legitimate arm's length transactions, involving actual physical liftings from the same location (Sullom Voe), between two partners disposing of the same volumes of the same blend in that location. They nicely illustrate the nature of tax spinning in its purest form.

The extent of outside trading (at least from Sullom Voe) seems to vary significantly from company to company. Outside trading, as opposed to intra-firm transfers, involves a higher proportion of BP's availabilities (89 per cent) than Shell's (69 per cent) or Esso's (50 per cent).

The 625,000 b/d of Sullom Voe blends acquired in 1984 by the three majors can be disaggregated by sources as follows:

internal transfers	:	201,000 b/d (32.2%)
cross-sales among the three	:	131,000 b/d (21.0%)
purchases from BNOC	:	136,000 b/d (21.8%)
purchases from others	:	157,000 b/d (25.1%)

The 588,000 b/d available to the three majors at Sullom Voe can be disaggregated by mode of disposal as follows:

internal transfers	:	201,000 b/d (34.1%)
cross-sales	:	131,000 b/d (22.2%)
other sales to Europe	:	113,000 b/d (19.2%)
other sales to North America	:	143,000 b/d (24.3%)

Thus, the three majors sold less than half the volume at their disposal to third parties (43.5 per cent). They also appear to have been significant customers of BNOC at Sullom Voe as they purchased 41.5 per cent of BNOC's availabilities.

CHAPTER 4

THE PATTERN OF INTERNATIONAL TRADE

4.1 Introduction

North Sea crudes are traded internationally in substantial volumes. A large share of North Sea production is exported, and the amounts traded have been increasing at high rates in recent years. These features of the North Sea warrant an analysis of international trade patterns, a task which we undertake in this chapter.

The following aspects shall be examined:

— the place of the UK and Norway in the world league of oil exporters
— the growth of gross and net exports of crude oil from the North Sea
— the delineation of the area in which North Sea crudes move in international trade
— the crude oil import structure of the main importing countries of North Sea crudes
— the contribution of North Sea oil to the balance of payments of the UK

The purposes of this analysis are to assess the significance of the North Sea in the world petroleum market, to identify the location of the North Sea export market, and to measure the degree of penetration of North Sea crudes in the main importing countries.

4.2 The UK and Norway as World Exporters

The importance of the UK and Norway relative to other oil-exporting countries may be established from Table 4.1, which ranks exporters in terms of their supplies to the world petroleum market. In 1984, for example, the UK exported some 1.06 mb/d of crude oil and 62,000 b/d of petroleum products (both figures net of imports). In that year only four OPEC countries exported a larger volume of crude and petroleum products than the UK – Saudi Arabia, Venezuela, Iran and Nigeria.

Table 4.1:   Ranking of Oil-exporting Countries by Export Volumes. 1984. Million Barrels per Day.

Volume	Countries
3.0 +	Saudi Arabia
2.0–less than 3.0	USSR
1.6–less than 2.0	–
1.1–less than 1.6	Mexico, Venezuela, Iran, Nigeria
0.9–less than 1.1	*UK*, UAE, Iraq, Kuwait, Libya
0.8–less than 0.9	Indonesia
0.7–less than 0.8	–
0.6–less than 0.7	–
0.5–less than 0.6	*Norway*, Algeria, Egypt
0.4–less than 0.5	–
less than 0.4	Oman, Qatar, Malaysia, Ecuador, Brunei, Angola, Gabon, Tunisia, etc.

Note: USSR figure inclusive of exports to other CPEs.
Source: Estimates derived from OPEC, *Annual Statistical Bulletin*, 1984; *Overseas Trade Statistics of the United Kingdom*, 1984; and *Uterikshandel*, 1984.

Two non-OPEC countries – Mexico and the Soviet Union – were larger exporters than the UK. On a world scale the UK would thus rank seventh.

Such a precise ranking is slightly misleading for two different reasons. First, the UK export volume in 1984 was close to those of Nigeria and the UAE. A distribution by brackets places the UK in the fourth rank (see Table 4.1). Secondly, the position of the UK would change significantly if world oil demand declined. A relatively small contraction would push the UK up the ladder because OPEC countries tend to absorb reductions in world oil demand in the performance of their role as residual suppliers. This has been happening in 1985.

Looking at the North Sea as a whole, and given recent world oil market developments, it is possible to state that in 1985 the North Sea ranked in third place in the league of world oil exporters after Saudi Arabia and the USSR.

4.3 Crude Oil Exports and Imports. The UK

According to the UK official trade statistics, *Overseas Trade Statistics of the United Kingdom* (various issues), the volume and value of gross exports of crude oil (ISIC 333: 'petroleum oils, crude and crude oils from bituminous materials') and the volume and value of imports were as shown in Table 4.2.

Table 4.2: UK. Exports and Imports of Crude Oil. 1981–4. Volume in Thousand Tonnes. Value in £ Millions.

Year	Exports		Imports		Net Exports	
	Volume	*Value*	*Volume*	*Value*	*Volume*	*Value*
1981	52549	7096	32990	4112	19559	2984
1982	60376	8497	28256	3951	32120	4546
1983	68279	10092	22803	3320	45476	6772
1984	75745	12228	24181	3847	51564	8381

Note: Net Exports = Exports minus Imports
Source: *Overseas Trade Statistics of the United Kingdom*, HMSO

The remarkable features of Table 4.2 are the growth in gross exports of crude oil between 1981 and 1984, the very significant size of crude oil imports, and the big decline in crude oil imports between 1981 and 1983 (1984 is an odd year because of the miners' strike).

It is interesting to compare changes in net exports with changes in crude oil production and petroleum consumption over the period under consideration. Table 4.3 shows much faster rises in the volume of net exports than in crude oil production.

Table 4.3: UK. Net Exports and Production of Crude Oil. Consumption of Petroleum Products. 1981–4. Million Tonnes. (Percentage Increases).

Year	Net Exports of Crude Oil	Production of Crude Oil	Total Petroleum Consumption
1981	19.6	87.9	74.8
1982	32.1 (64%)	100.3 (14%)	75.6 (1.1%)
1983	45.5 (42%)	110.5 (10%)	72.4(−4.2%)
1984	51.6 (13%)	120.8 (9%)	89.6 (23.8%)

Data on petroleum trade can also be obtained from OECD, *Oil and Gas Statistics*, a convenient source for inter-country comparisons because the figures are presented in a consistent form for all OECD member countries. But the OECD data are not strictly comparable with UK national data because of definitional differences. The OECD source adopts a wider concept than ISIC 333 as it includes crude oil, natural gas liquids (NGLs) and refinery feedstocks under the same item. Table 4.4 presents the OECD figures for UK exports and imports for the period 1981–4.

Table 4.4:   UK. Exports and Imports of Crude Oil, NGLs and Refinery Feedstocks. 1981–4. Thousand Tonnes.

Year	Exports	Imports	Net Exports
1981	52180	36827	15353
1982	61645	33753	27892
1983	69930	30323	39607
1984	80360	32271	48089

Source: OECD, *Oil and Gas Statistics*

4.4   Crude Oil Exports and Imports. Norway

Data on oil trade for Norway can similarly be obtained from:

— the trade publication of the Central Bureau of Statistics of Norway, *Uterikshandel*
— OECD, *Oil and Gas Statistics*

The first source adopts the narrow ISIC 333 concept and is therefore comparable with *Overseas Trade Statistics of the United Kingdom*. Export and import data from this source are presented in Table 4.5.

According to this source, gross export volumes remained virtually unchanged between 1981 and 1982, but volume increased in 1983 by 24 per cent and in 1984 by 17 per cent. Norway differs from the UK in one important respect: crude oil imports are small relative to the country's export level. Nevertheless, Norway shares with the UK a steady decline in the volume of crude oil imports between 1981 and 1983.

Table 4.5: Norway. Exports and Imports of Crude Oil. 1981–4. Volume in Thousand Tonnes. Value in Kr Millions.

Year	Exports		Imports		Net Exports	
	Volume	Value	Volume	Value	Volume	Value
1981	20453	31047	3200	6054	16553	24993
1982	20666	31878	2927	4668	17739	27210
1983	25623	40653	1758	2901	23865	37752
1984	29972	52278	2025	3487	27947	48791

Source: *Uterikshandel*

The OECD source, *Oil and Gas Statistics*, gives a different picture of Norway's oil trade (see Table 4.6).

Table 4.6: Norway. Exports and Imports of Crude Oil, NGLs and Refinery Feedstocks. 1981–4. Thousand Tonnes.

Year	Exports	Imports	Net Exports
1981	18871	2908	15963
1982	19230	2778	16452
1983	26058	1896	24162
1984	29073	2279	26794

Source: OECD, *Oil and Gas Statistics*

The discrepancies relate to a peculiarity in the transport of Norwegian oil from field to terminal. Norwegian crudes from Ekofisk and Murchison (Norway) are shifted directly by pipeline to the UK terminals at Teesside and Sullom Voe respectively. The Central Bureau of Statistics of Norway seems to treat the *whole* of the volume so shifted as exports to the UK without netting out from these deliveries the amounts (small as they may be) that are returned to Norway. These amounts are treated as imports into Norway and added to the import figures. This approach inflates both oil export and import figures.

Treating all pipeline deliveries to Teesside and Sullom Voe as exports to the UK conceals an important fact, namely that a substantial proportion of Norwegian oil is exported from the UK to other destinations. Thus the geographical distribution of crude oil exports derived from Norway's statistical source, *Uterikshandel*, is grossly distorted, showing the UK to be a much larger importer of Norwegian crude, and other countries to be much smaller importers, than they actually are.

4.5   The Destination of North Sea Oil Exports

The purpose of this section is to identify which countries are the major importers of North Sea crudes and to determine their rank and their import shares. The analysis of trade data by destination will enable us to delineate the geographical boundaries of the international market for North Sea crudes and to assess the degree of geographical spread.

(a) The UK. The geographical distribution of UK oil exports (crude plus NGLs, plus refinery feedstocks as defined by OECD) is shown in Table 4.7, which presents data provided by OECD, *Oil and Gas Statistics* under the heading 'Country X Imports from the UK'.

The ranking of importing countries by volume puts the USA in first place, and Germany in second, in each of the four years 1981–4 (see Table 4.7). The Netherlands occupies third or fourth place. France's imports have increased very substantially every year; consequently France has moved from seventh place in 1981 to third in 1983 and 1984. Sweden's import volume has risen significantly in 1983 and 1984.

The degree of geographical concentration of UK oil exports is high. In the period 1981–4, the four countries at the top of the list (the USA and three major European importers) accounted for 73–83 per cent of UK exports; and the six major importers for about 87–92 per cent. Although there is a decline in the degree of concentration over time, the geographical spread of UK crude oil exports (given that six countries account for almost 90 per cent of the total volume) is very narrow. UK oil exports flow mainly to the USA and North West Europe. The countries which have proved important markets for UK oil include some of the major oil importers (USA, Germany, France); but several others (Japan, Italy, Brazil, Spain) do not participate in any significant way in North Sea trade.

(b) Norway. Norway exports its crudes to a similar set of countries to the UK, but there is an important difference. The UK is itself a major importer of Norwegian crudes, occupying first place in certain years. Another difference is that neither Denmark nor Belgium is an importer.

Table 4.7: Main Importing Countries of UK Crude Oil. 1981–4. Thousand
Tonnes.

Country	Imports	% of Total
1981		
1. USA	18208	34.3
2. Germany	15947	30.0
3. Netherlands	4144	7.8
4. Sweden	3280	6.2
5. Belgium	2904	5.5
6. Denmark	2450	4.6
7. France	2336	4.4
1982		
1. USA	22015	36.8
2. Germany	15353	25.7
3. Netherlands	7659	12.8
4. France	4463	7.5
5. Sweden	3555	5.9
6. Denmark	2119	3.5
7. Norway	1463	2.4
1983		
1. USA	18070	26.9
2. Germany	14301	21.3
3. France	8968	13.3
4. Netherlands	7618	11.3
5. Sweden	5943	8.8
6. Belgium	4269	6.3
7. Denmark	2619	3.9
1984		
1. USA	18575	23.7
2. Germany	17808	22.7
3. France	13024	16.6
4. Netherlands	7818	10.0
5. Sweden	6372	8.1
6. Belgium	4658	5.9
7. Denmark	2213	2.8

Source: OECD, *Oil and Gas Statistics*

Table 4.8:   Main Importing Countries of Norway's Crude Oil. 1981–4. Thousand
Tonnes

Country	Imports	% of Total
1981		
1. USA	5598	31.3
2. UK	3126	17.5
3. Germany	2895	16.2
4. France	1876	10.5
5. Netherlands	1398	7.8
6. Sweden	1086	6.1
1982		
1. USA	4826	26.1
2. UK	3533	19.1
3. Sweden	2733	14.8
4. France	2508	13.6
5. Germany	2432	13.1
6. Netherlands	1787	9.7
1983		
1. UK	6943	29.4
2. Germany	3804	16.2
3. Sweden	3129	13.3
4. USA	3078	13.1
5. Netherlands	3048	13.0
6. France	1872	8.0
1984		
1. UK	9504	35.7
2. USA	5284	19.9
3. Netherlands	3566	13.4
4. Sweden	3119	11.7
5. Germany	2614	9.8
6. France	1858	8.0

Source: OECD, *Oil and Gas Statistics*

The six major importing countries are the USA, the UK,
Germany, Sweden, France and the Netherlands. The ranking
changes from year to year; but the countries in the set remain the
same. The difference with the UK is that for Norway the number of
small importers is not very large.

There seems to be a decline in French imports of Norwegian oil and a steady increase in the imports into the UK and the Netherlands. In other cases the volumes tend to fluctuate.

The degree of geographical concentration is high, for the reason mentioned above. The six main importing countries accounted for 97.5 per cent of total crude oil exports in 1984.

4.6 North Sea Crudes *vis-à-vis* Other Crudes in the Major Importing Countries

The purpose of this section is to determine the place of North Sea crudes in the imports of the seven countries identified in Table 4.7 and 4.8 as major importers of UK and Norwegian oil. We are particularly interested in the competition with crude oil imports from other exporters of the Atlantic Basin. For the purposes of this section the oil-exporting countries of the Atlantic Basin are taken to include:

North Sea	:	UK and Norway
North America	:	Canada and Mexico
South America	:	Venezuela
Africa	:	Algeria, Libya and Nigeria

We shall ignore the many smaller exporting countries of the Basin (Angola, Tunisia, Trinidad, Gabon, etc.), partly because their impact is small and partly because trade statistics are not sufficiently disaggregated.

The picture drawn in Table 4.9 reveals interesting and significant differences between the major importing countries of North Sea oil. First, the import dependence on North Sea crudes seems to vary considerably from country to country. Sweden appears to be strongly oriented towards the North Sea (some 70 per cent of its 1984 crude oil imports came from the UK and Norway). The USA is at the other end of the spectrum. The USA is the largest importer of UK crudes and one of Norway's largest oil customers, but these countries' shares in the US import bill is small (11.5 per cent). Neither the UK nor Norway is a very significant supplier of crude oil to the mammoth US economy.

The North Sea share of oil imports for the five other countries (Germany, France, the Netherlands, Belgium and Denmark) ranges between 20 and 38 per cent. These shares are substantial, and they increased significantly between 1983 and 1984.

Table 4.9: Crude Oil Imports of Seven Countries. 1984. Thousand Tonnes.

Exporters	USA	Germany	France	Netherlands	Sweden	Belgium	Denmark
				Importing Countries			
UK	18575	17808	13024	7818	6372	4658	2213
Norway	5284	2614	1858	3566	3119	383	–
Total North Sea	23859	20422	14882	11384	9491	5041	2213
Canada	24549	–	–	–	–	–	–
Mexico	37610	2	3563	3708	–	–	–
Venezuela	15256	4210	770	322	904	566	–
Algeria	9496	2671	5237	2521	115	80	–
Libya	–	9636	3550	1812	–	1468	–
Nigeria	10547	9530	9512	3546	499	2879	247
Total Other Atlantic	97458	26049	22632	11909	1518	4993	247
Total Atlantic Basin	121317	46471	37514	23293	11009	10034	2460
Total Crude Oil Imports	206696	66933	73505	44743	13426	23417	5818
			Shares of Total Crude Oil Imports. Percentages.				
UK	9.0	26.6	17.7	17.4	47.5	19.9	38.0
Norway	2.5	3.9	2.5	8.0	23.2	1.6	–
North Sea	11.5	30.5	20.2	25.4	70.7	21.5	38.0
Other Atlantic	47.2	38.9	30.8	26.6	11.3	21.3	4.3
Atlantic Basin	58.7	69.4	51.0	52.0	82.0	42.8	42.3

Source: OECD, *Oil and Gas Statistics*

The USA depends more on other Atlantic suppliers (mainly Mexico for obvious reasons of proximity, but also Canada, Venezuela and Nigeria) than on the North Sea. The import share of other Atlantic exporters was 47.2 per cent in 1984 compared with 11.5 per cent for the North Sea.

In 1984, France imported more oil from the UK than from any other source. Germany was a large importer of Libyan (9.636 mt) and of Nigerian oil (9.530 mt), and a substantial one of Venezuelan crudes (4.210 mt). The share of 'other Atlantic exporters' was 39 per cent and that of the North Sea 30.5 per cent.

Swedish and Danish imports display totally different characteristics. As mentioned above, Sweden is North Sea orientated while Denmark imports only 38 per cent of its oil from the North Sea and very little (4.3 per cent) from other Atlantic exporters.

Finally, the dependence of these importing countries on the Atlantic Basin, including the North Sea, is high. The Atlantic Basin share is over 80 per cent in Sweden, close to 70 per cent in Germany and close to 60 per cent in the USA. It is just over 50 per cent in France, and around 42 per cent in Belgium and Denmark.

But these large 'Atlantic' shares should not conceal the fact that Gulf exporters still supply a very large proportion of the oil imports of major Atlantic consumers. Furthermore, the import dependence of these consumers (except Sweden) on OPEC is even higher because several Atlantic exporters (Venezuela, Algeria, Libya, Nigeria) are OPEC members. In fact, the oil import dependence on OPEC is as high as 78 per cent in Japan (a country which does not import North Sea oil because of the distance involved) and 40–70 per cent in the other main industrial countries.

OPEC shares compared with those of the North Sea and other non-OPEC producers are shown in Table 4.10. The share of OPEC imports has declined between 1983 and 1984 in some European countries – particularly Germany, France, the Netherlands and Belgium, but did not change in the USA and Japan. In Europe the reduction in OPEC's share is not entirely attributable to the North Sea. The Soviet Union and other non-OPEC exporters have all expanded their shares relative to OPEC.

4.7 UK. Oil and the Balance of Payments

Between 1981 and 1984 the value of net crude oil exports increased substantially from approximately £3 billion to £8.4 billion (180 per cent). The contribution of crude oil trade to the UK balance of

Table 4.10: North Sea, Other non-OPEC and OPEC Shares in Crude Oil Imports of Major OECD Consuming Countries. 1983–4. Percentages.

Exporters				*Importing Countries*				
1983	USA	Japan	Germany	France	Italy	Spain	Netherlands	Belgium
North Sea	10.7	–	27.7	15.1	1.3	3.6	26.5	22.6
Other non-OPEC	47.3	22.0	11.2	21.1	28.8	31.3	18.9	30.0
(of which CPEs)	(0.3)	(5.3)	(6.8)	(8.7)	(12.4)	(1.8)	(5.7)	(15.1)
OPEC	42.0	78.0	61.1	63.8	69.9	65.1	54.6	47.4
1984								
North Sea	11.5	–	30.5	20.2	2.0	3.2	25.4	21.2
Other non-OPEC	46.5	22.3	13.9	21.8	29.6	32.8	24.4	37.7
(of which CPEs)	(0.1)	(6.1)	(8.6)	(7.5)	(15.0)	(3.0)	(9.0)	(24.3)
OPEC	42.0	77.7	55.6	58.0	68.4	64.0	50.2	41.1

Source: OECD, *Oil and Gas Statistics*

payments thus became very considerable, both in absolute terms and relative to other trade aggregates (see Table 4.11).

Table 4.11: UK. Crude Oil Exports and the Balance of Payments. 1981–4. Value in £ Millions.

Year	UK Exports Goods & Services	Current Account Balance	Net Crude Oil Exports	Crude Oil Exports as % of	
				UK Exports	Current Balance
1981	68042	+6929	2984	4	43
1982	73190	+4934	4546	6	92
1983	80077	+2543	6772	8	266
1984	91652	+51	8381	9	16433

Sources:*Economic Trends* and Table 4.2 above.

While the share of net crude oil exports in total UK exports of goods and services, though rising, is small (4 per cent in 1981 but increasing rapidly to 9 per cent in 1984), the value of net crude oil exports relative to the current balance-of-payments surplus is large. In 1981 crude oil exports, which were then fairly modest, accounted for 43 per cent of the UK balance-of-payments surplus, then at its largest.

In 1984 the balance on the UK current account was virtually nil. Assuming other things remained equal, a very small change in net oil export revenues would put the balance in the red. A 10 per cent decline in the price of oil would have no noticeable effect on the UK export earnings from goods and services, but would have a serious impact on the current balance of payments.

4.8 Conclusions and Implications

The international market for North Sea crudes consists of North West Europe and the USA. Rapid export growth, resulting from the expansion of production and the decline in UK oil consumption, has secured large shares for North Sea crudes in the markets of major importing countries. In 1985 the North Sea, as a whole, has become one of the two or three largest suppliers of internationally traded oil.

There is an interface in Europe between the North Sea and the main Gulf and African crudes; and an interface in the USA with WTI and imported Atlantic Basin crudes. In Europe, North Sea crudes are short-haul oil. This means that changes in European market conditions have an immediate impact on North Sea spot prices. In the USA, the relationship between spot prices of North Sea and US crudes (such as WTI) is less direct. The more direct price links are between US oil futures and North Sea forward prices.

It is important to recall that though the USA is the largest importer of North Sea crudes, the North Sea only commands a small share of the US market. The low dependence of the USA on the North Sea explains situations, rare as they may have been, in which the two markets moved apart and prices ceased to relate closely. This happened, for example, in December 1984/January 1985.

It can be seen that the analysis of trade patterns raises questions about market behaviour and prices which are taken up in subsequent parts of this book, and at the same time provides some of the elements for an answer.

CHAPTER 5

THE COSTS AND TIME PATTERNS OF PRODUCTION

5.1 Introduction

The approach we have adopted for studying the market for North Sea crude oil started by identifying the commodities traded and their characteristics. The next step focused on the various agents operating on the supply and end-user sides of the market, and provided a description of the market structure and measures of the degree of concentration. Since North Sea crudes are traded internationally, it was necessary to define the boundaries of the geographical area in which these crudes move physically from sources of production to refineries, and to study the pattern and growth of exports.

These various steps described aspects of the market setting in which the forces of supply and demand operate. The purpose of this chapter is to analyse in some detail two issues relating to supply. These are:

— the costs of production in North Sea oilfields
— the time pattern of oil production in the North Sea

Costs are usually a determinant of the output policies of the firm. Furthermore, the analysis of costs will shed light on an issue which has recently become topical: the impact of a fall in the price of oil on North Sea oil supplies.

The temporal production pattern of North Sea oil is studied because it involves both growth and seasonal fluctuations. The growth of North Sea oil output obtained in a period during which world demand for petroleum was depressed; it was part of a general expansion of non-OPEC production which contributed to a fall in the nominal price of oil by some $10 between 1980 and 1985. The fluctuations of North Sea production raised questions about their causes. The general position of the oil industry on this issue, that output fluctuations are entirely due to technical factors, such as maintenance, is not universally shared by others. Economists tend to believe that economic factors such as prices have an influence on output decisions. Others think that government policy has a role to play. As recently as June 1985 some observers speculated that maintenance programmes were being brought forward in order to stabilize spot prices on the eve of the OPEC meeting of 5th July.[1] A thorough examination of the possible causes of oil output fluctuations is undertaken here to clarify this controversial issue.

5.2 Costs of Production

(a) Costs and Output Maximization. The exploration and development of North Sea oilfields involves substantial capital costs; but contrary to the views of certain ill-informed observers outside the industry, the average operating costs of all producing fields are low relative to the current price of oil. However, both capital and operating costs are intimately linked with decisions on how much oil will be produced from the North Sea.

The general principle facing any company with a need for crude oil is that as much as possible should be acquired from the source that yields the biggest margin between tax-paid cost and revenue; then if this source is insufficient for total needs recourse should be made to the source with the second largest margin, and so on.

For companies that both produce and use crude there is an important distinction between the fields they own for which the allowances for development costs have been fully used and those for which they are still able to claim tax relief. The oil fiscal regimes in both the UK and Norway allow for the cumulative deduction of capital costs (with an uplift) from revenues before the payment of certain taxes. This system ensures that the margin is much greater on oil from non-tax paying fields than from any other source, and hence that crude from these fields will be fully utilised before other sources of supply are brought into operation. (This will be subject of course to the usual constraints imposed by good field management and by government regulations on best-practice techniques, gas flaring and other relevant factors.) Indeed the North Sea tax systems, compared with the familiar alternative which provides for a depreciation allowance over the notional lifetime of the assets, tend to shorten the period over which capital expenditures are recouped.

[1] In fact maintenance work in the UK sector started later in 1985 than in either 1983 or 1984.

For fields which have become fully tax paying the margin is clearly smaller since there are no capital-related tax offsets to claim. In the present conditions of the oil market the North Sea cost structure and fiscal regimes provide firms with relatively high margins compared to other sources, so the firms will tend to take as much from these fields as possible.

However, were the price of oil to fall, the use of the North Sea as a priority source could conceivably change for fields that have become fully tax paying while it would not change for those that have not. An alternative source of crude, having different operating costs and a different tax structure, could switch from having a lower margin to having a higher margin at a price well above the operating costs of the North Sea fields.[2]

This general tendency to extract oil at maximum rate does not necessarily rule out occasional exceptions dictated by short-term economic or political factors. Furthermore economists believe that expectations of a future price rise can lead to output restrictions depending on rates of discount and the relevant time horizon. Whether oil companies take these expectations into account is a moot point. However, what is certain today is that nobody expects oil prices to rise in the medium term; and for the moment the optimum depletion argument is of little relevance.

(b) Operating costs. Estimates of operating costs per barrel for 1984 are presented in Table 5.1. The table covers all producing fields in the UKCS and Norway. The estimates are obtained by dividing total operating costs as provided in Wood Mackenzie (the only easily available source of cost data) by the annual output of each field in 1984.

The lowest operating cost per barrel of production is $1.3 (Forties) and the highest is $12.5 (Tartan). The range of operating costs appears at first sight to be wide. However, all the big North Sea fields, except Ekofisk, operate at very low costs ($1.3–3.3), and the proportion of North Sea oil output involving high operating costs is not very large. Table 5.2 shows the 1984 volumes and proportions of North Sea oil output at cut-off points determined by the operating costs of production. Two alternative calculations are presented:

[2] The introduction of netback deals and the fall in oil prices in late 1985 have together brought about the conditions such that this is no longer merely a theoretical possibility. Indeed an oil company might now realistically expect to secure a better per barrel margin from certain OPEC deals than from some North Sea producing fields.

Table 5.1:   UKCS and Norway. Operating Costs per Barrel by Producing Oilfield. 1984. Dollars.

Field	Cost	Field	Cost
UKCS			
1. Argyll	7.5	16. Magnus	3.5
2. Auk	10.7	17. Maureen	3.3
3. Beatrice	5.9	18. Montrose	8.0
4. Beryl	4.9	19. Murchison	2.0
5. Brae	4.4	20. Ninian	3.3
6. Brent	2.2	21. Piper	1.9
7. Buchan	6.9	22. Statfjord	1.7
8. Claymore	2.5	23. Tartan	12.5
9. Cormorant	3.4	24. Thistle	3.4
10. Duncan	9.6		
11. Dunlin	3.2	*Norway*	
12. Forties	1.3	1. Ekofisk (area)	5.4
13. Fulmar	1.6	2. Murchison	2.0
14. Heather	6.2	3. Statfjord	1.6
15. NW Hutton	3.5	4. Valhall	8.2

Note: Calculations based on total operating costs (both oil and gas) as they cannot be separated.
The actual costs of oil production in fields that produce both oil and gas are lower than shown.

Source: Wood Mackenzie

— The first assumes that companies will continue to produce from a field until the average operating cost on the whole output of the fields becomes equal to the price. (The average cost is taken to be the same as the marginal.) This can only happen *without losses* if the Government ceases to levy taxes, including royalties, and if capital expenditures have been fully amortised. There are many situations where firms continue to produce at a loss at least for a while because capital costs, whether amortised or not, are sunk costs.

— The second set of calculations assumes that the Government does not waive its claim on royalties. In this case, per barrel operating costs are inflated by a coefficient representing the royalty payment made at present at an average actual rate of approximately 11 per cent.

Table 5.2 suggests that a fall in the price of oil to as low as $5 would not reduce North Sea oil production, in the short term, by more than 7 per cent. The actual impact on production of such a price decline would probably be greater than 7 per cent because

Table 5.2: Proportion of North Sea Output Likely to be Shut Down at Various Oil Prices.

Oil Price = Operating Costs per barrel (excl Royalties) $	Oil Price = Operating Costs per barrel (incl Royalties) $	Proportion of Total Production Lost at Different Prices %		
		North Sea	UKCS	Norway
12.5 (+)	14.0 (+)	nil	nil	nil
10.7	12.0	0.9	1.1	nil
9.6	10.8	1.3	1.6	nil
8.2	9.2	1.6	2.0	nil
8.0	9.0	3.2	2.0	7.2
7.5	8.4	3.7	2.7	7.2
6.9	7.8	4.0	3.0	7.2
6.2	7.0	4.6	3.8	7.2
5.9	6.6	5.4	4.8	7.2
5.4	6.1	6.8	6.7	7.2
4.9	5.5	14.8	6.7	43.1
4.4	4.9	17.6	10.2	43.1
3.5	3.9	20.3	13.7	43.1
3.4	3.8	25.3	20.1	43.1
3.3	3.7	32.0	28.7	43.1
3.2	3.6	41.8	41.3	43.1
2.5	2.8	43.8	43.9	43.1
2.2	2.5	46.9	47.9	43.1
2.0	2.2	60.1	65.0	43.1
1.9	2.1	63.4	68.2	46.8
1.7	1.9	69.0	75.5	46.8
1.6	1.8	71.3	78.4	46.8
1.3	1.5	87.1	83.4	100.0
1.3 (−)	1.5 (−)	100.0	100.0	100.0

certain companies might find themselves in serious financial difficulties, and others might find it more rewarding to obtain oil from other sources. The significant conclusion is that the short-term supply elasticity is very small. The more important effects of a price fall are those on long-term production, on producing governments' tax revenues, and on the financial institutions that have lent funds to oil-exporting countries or to energy corporations.

5.3 North Sea. Total Oil Production Pattern

North Sea oil production displays two important features:

— steady growth over the years
— marked fluctuations within the year

Both characteristics have an influence on the behaviour of the market. Furthermore, seasonal fluctuations raise interesting questions about their causes. The time patterns of UKCS and Norwegian output will now be examined in turn.

Figure 5.1 shows the steady growth in UKCS oil production from January 1980 to December 1985. The data sample, taken from statistics published by Wood Mackenzie, therefore contains seventy-two observations. The annual average growth rate of UKCS oil production during that period is estimated at 10.15 per cent, and the equivalent monthly growth rate is 0.809 per cent.

Casual observation of the graph would suggest a fairly *smooth* upward trend until the beginning of 1983. Beyond this point UKCS oil production continues to increase on trend but appears to become increasingly prone to cyclical variations. In both 1983 and 1984 peaks occur in February and troughs in June. In 1985 peak production occurs in January but the trough remains in June.

Statistical analysis, presented in detail in Annex 1 for the period January 1980 to February 1985, enables us to measure a linear trend and test for the seasonality of UKCS oil production. The coefficients of the TREND variable are always significant, and indicate a monthly trend increase in UKCS production between January 1980 and February 1985 of 19,500 b/d, almost a quarter of a million b/d in a year.

There is also evidence of seasonality in UKCS oil production. The statistical analysis undertaken, using regressions, shows that the winter quarter (December, January, February) is associated with an average rise in production of 60,500 b/d relative to the autumn quarter. In contrast, the summer quarter sees an output fall of 61,500 b/d relative to the autumn norm.

These seasonal variations are not insubstantial compared to the mean of the data sample (2.083 mb/d). They are equivalent to a change in output of approximately 3 per cent. The seasonal swings of plus or minus 60,000 b/d, in the relevant months, may also be compared with the monthly trend increase of 19,500 b/d.

Turning now to the Norwegian case, Figure 5.2 depicts the slight upward trend in total Norwegian oil production over the data period (January 1980–December 1985). Around the trend there is considerable fluctuation but this does not seem to occur on any regular basis. Of particular note are the large output falls in July 1980 (287,000 b/d), August 1981 (203,000 b/d) and August–October 1983 (92,000 b/d).

Over the period January 1980–December 1985, the monthly (annual) growth rate was estimated to be 0.813 (10.21) per cent.

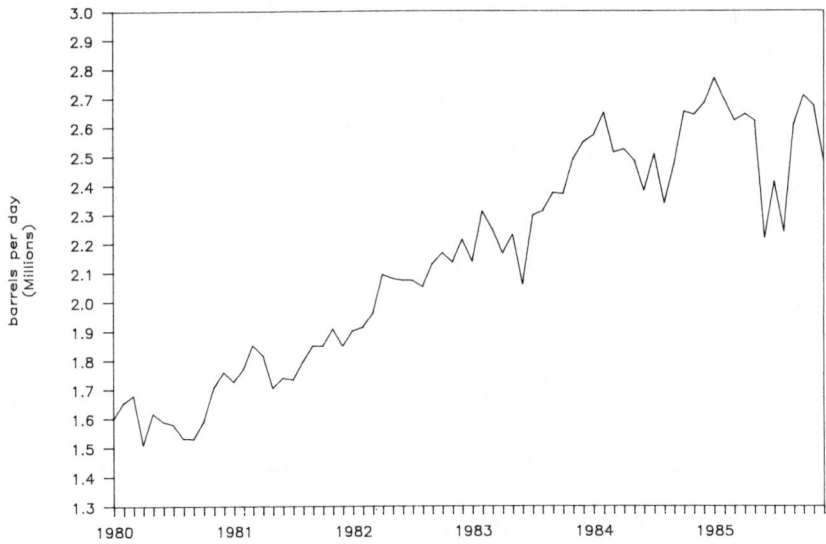

Figure 5.1 UKCS. Total Monthly Oil Production 1980–1985

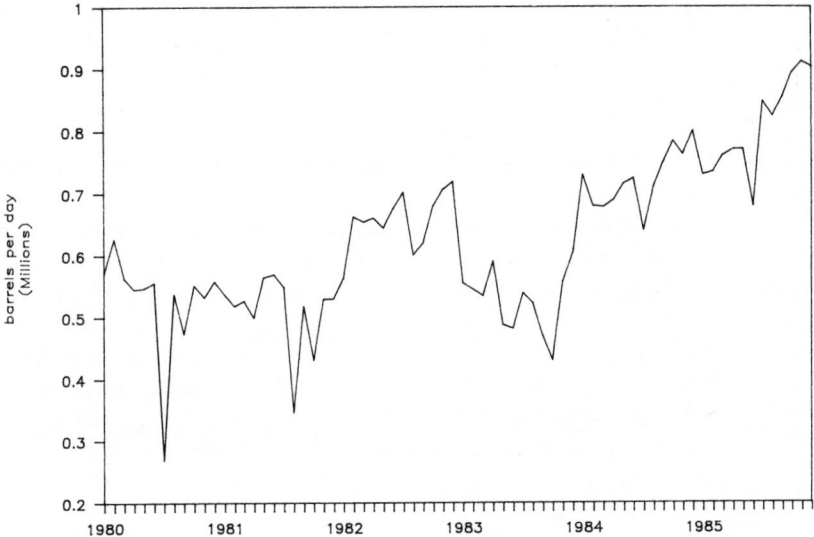

Figure 5.2 Norway. Total Monthly Oil Production 1980–1985

The more detailed statistical analysis for the shorter period (January 1980–February 1985) provides only weak evidence of output seasonality. Average output increases are 59,500 and 58,100 b/d in the winter and spring respectively, relative to the autumn quarter (August, September, October). Overall, the results indicate a low output level in late summer/early autumn relative to the rest of the year.

We also tested for seasonality in production for certain major UKCS and Norwegian fields considered separately. However, since neither technical nor economic explanations of seasonality for fields in the Norwegian sector derive much support from our statistical analysis, we confine our attention in this chapter to our results for selected UKCS fields as presented below.

5.4 UKCS. The Production Pattern of Selected UKCS Oilfields.

The same analysis is now undertaken for each of the major UKCS oilfields: Brent, Forties, Ninian and Piper. The other fields are aggregated together and the production pattern of this Rest of the UKCS is assessed in the same way. The period covered in all cases remains January 1980 to December 1985 in the Figures, and January 1980 to February 1985 in the statistical computations.

(a) The Brent field. Actual production of the Brent field rose from 126,000 b/d in January 1980 to 467,000 b/d in December 1985. The peak monthly output of this period was 518,000 b/d attained in February 1984. The 1980 average output of the Brent field was 140,000 b/d, to be compared with an average production of 424,000 b/d in 1985. The production time-profile for the Brent field is shown in Figure 5.3.

The strong upward trend is apparent from the graph. The outstanding feature, however, is the large fluctuations in production, particularly noticeable in 1983 and 1984. In 1983 and 1984, February is a peak month followed by four months of successive decline – a dramatic swing of 260,000 b/d in 1983, and 199,000 b/d in 1984. In both years the trough occurred in June. In 1985 output was lower in February than in January but this slight difference from the pattern observed in 1983 and 1984 is of no significance.

The more interesting observations are that:

— Average production was lower during the eight months follow-

ing the June 1984 trough than in the comparable eight months
following the June 1983 trough.
— There is no rising trend over the four consecutive output peaks
of December 1982 (492,000 b/d), February 1983 (480,000 b/d),
February 1984 (518,000 b/d) and January 1985 (486,000 b/d).
All these peaks are just below or slightly above the 500,000 b/d
mark.

These observations suggest that the oil output of the Brent field
may well have reached a plateau, at least under current recovery
techniques.

The statistical analysis of the monthly output data of Brent
confirms the existence of significant seasonality. The spring and
summer quarters are associated with average production falls of
51,500 and 41,400 b/d respectively, relative to the autumn level.
The significance of the production drop in these quarters is better
appreciated when compared with the mean of the data sample
(309,000 b/d).

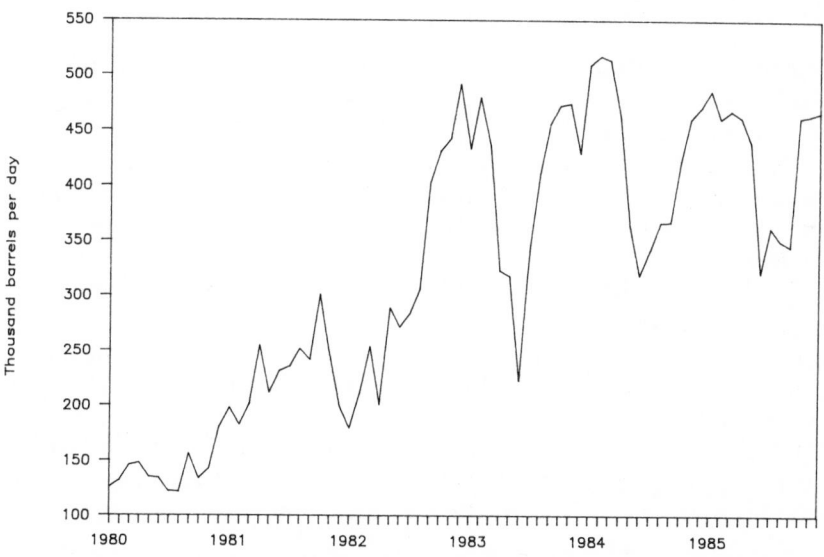

Figure 5.3 UKCS. Brent Oilfield. Monthly Oil Production 1980–1985

(b) The Forties field. Figure 5.4, which plots the monthly output of
Forties, suggests a slight downward trend around which there

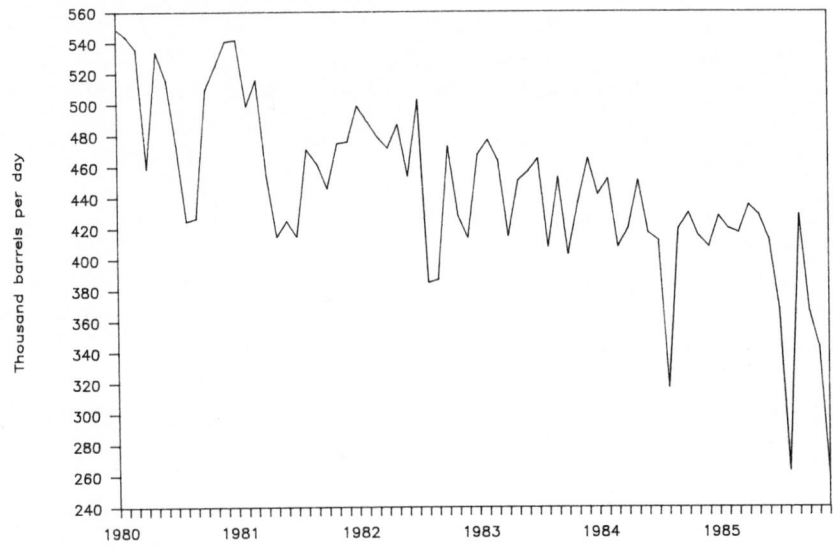

Figure 5.4   UKCS. Forties Oilfield. Monthly Oil Production 1980–1985

appear to be significant fluctuations. The trend in Forties' oil production clearly indicates that Forties has already peaked and is on a gentle decline.

The notable features of the graph are the large drop in production in August 1984 (around 100,000 b/d), caused by a brief shut-down to tie in the Montrose pipeline, and the subsequent recovery.

There is a statistically significant trend decrease in production equivalent to a monthly average of 1700 b/d. The seasonality is also confirmed by the coefficients of the relevant seasonal variables. Output rises on average at the rate of 18,700 b/d in the winter and falls by 33,100 b/d in the summer relative to the autumn.

However, these movements are less sharp than those for Brent, since their magnitude is not as large relative to the mean of the data sample (456,000 b/d) as in the case of Brent.

(c) The Ninian field. The time-profile of oil production for the Ninian field depicted in Figure 5.5 does not show an overall trend. Output generally tended to increase until mid-1982 and began to decrease thereafter. Output fluctuations seem to have been more marked since January 1984 than in earlier years.

The statistical analysis failed to reveal regularities in the temporal pattern of Ninian production. Splitting the data into two periods (January 1980–June 1982 and July 1982–February 1985) confirms the change in trend that appears from casual observation. In the first period the strong upward trend is equivalent to a monthly trend increase of 3900 b/d, and in the second period output declines rapidly at a monthly trend rate of 3100 b/d.

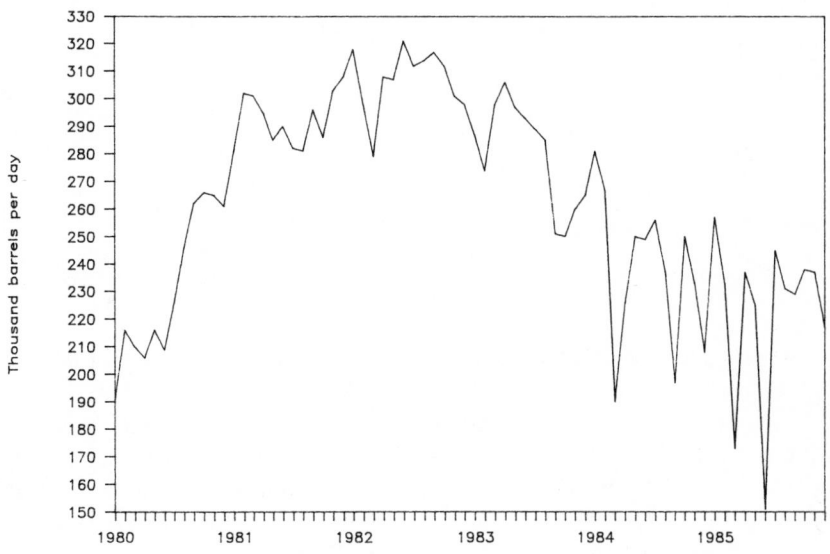

Figure 5.5   UKCS. Ninian Oilfield. Monthly Oil Production 1980–1985

(d) The Piper field. The notable feature of the Piper field is the remarkably constant production over much of the data period, as can be readily seen from a glance at Figure 5.6. There are no consistent signs of seasonality. The only significant fluctuations occurred between March and June of 1984. Output dropped by 100,000 b/d between February and April 1984; it recovered in May, fell back in June and finally moved up close to the period average (199,000 b/d). It would appear that some special factor (in fact, a fire) was operating during these perturbed months. Otherwise oil production from Piper was not subject to significant seasonal variations. So far Piper appears to have been a remarkably steady field.

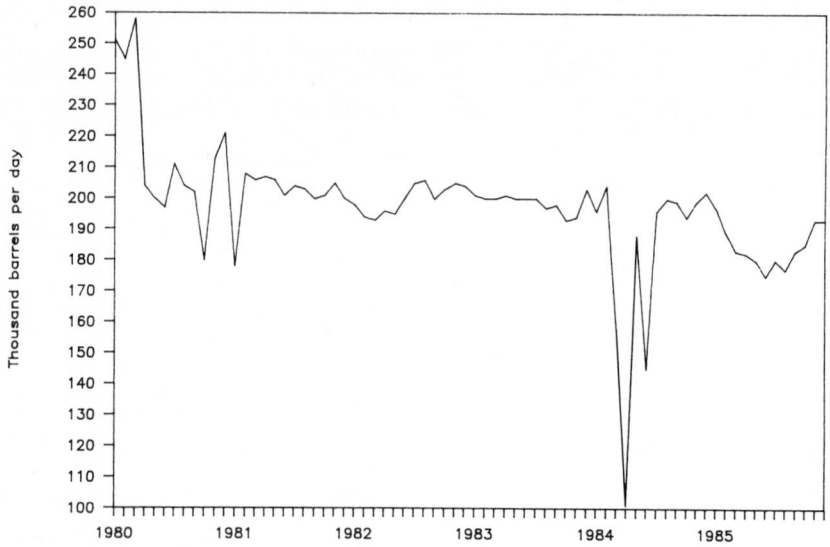

Figure 5.6   UKCS. Piper Oilfield. Monthly Oil Production 1980–1985

(e) The Fulmar field. This field began operating in February 1982: thus the data base includes fewer observations than those for other fields. There is weak evidence of seasonality in Fulmar. The statistical analysis indicates an average increase of 27,000 b/d in the winter relative to the reference season (autumn).

(f) The Claymore field. The statistical analysis shows that the TREND variable is significant and indicates a monthly increase of about 200 b/d. However, the statistical analysis fails to reveal any seasonal fluctuations.

(g) The Dunlin field. The statistical analysis reveals seasonality and the results are established at a high level of confidence. There is a small trend decline equivalent to a monthly average drop of 900 b/d. In contrast the seasonal increases are in the range of 14,200–19,100 b/d in the relevant quarters.

(h) The Cormorant fields. The interesting feature of the production time-profile of Cormorant is the relatively long shut-down of May–September 1981 which brought output down to zero for five

months. The statistical analysis shows some evidence of seasonality, but this may have something to do with the long shut-down of 1981 and a subsequent shut-down in April 1982.

(i) The Rest of UKCS. The Rest of UKCS is defined here as the difference between total crude oil production and the output of four major fields already considered (Brent, Forties, Ninian and Piper). The statistical analysis shows that the trend is equivalent to an average monthly increase in output of 15,800 b/d, but does not yield any conclusive evidence of seasonality except for a weak movement in the spring.

5.5 Causes of Output Variations in UKCS

There are a number of reasons why production might vary during the year and between years. Some of the most likely include:

— changes in the physical behaviour of existing oilfields and emergence of new fields
— weather-related effects in offshore loading fields
— maintenance-related effects
— economic choice

If oil producers in the UKCS behaved consistently and continually as output maximizers, economic factors, such as price or demand variations, would not cause changes in the production time-profile. The economic decision to maximize output, subject to technical constraints, would be taken once and for all. Any actual changes in the output time-profile would be entirely due to changes in effective capacity brought about by the technical, maintenance and other non-economic factors mentioned above.

Let us now examine these factors in turn and assess whether they can explain seasonality. The first factor, which may be succinctly referred to as 'changes in maximum flows', covers variations in effective capacity which obtain over time as new fields start up and old fields are run down. It also includes changes or discontinuities in the physical behaviour of fields such as a drop in pressure or a sudden gush. These rather erratic occurrences are distinct from the smooth decline in production of an old field that has passed its peak.

Over a period of time changes in maximum flows should yield a fairly smooth adjustment when production is aggregated over many fields. Erratic behaviour on the part of some fields might cause some sharp month-to-month variations, but it is difficult to see how these factors would explain a regular seasonal pattern of output changes involving large fluctuations in, say, the summer or the winter.

The second factor, which refers to weather-related effects on offshore loading fields, would provide a plausible explanation of seasonality if both:

— A large proportion of UKCS output was loaded off shore.
— Peak production occurred during the good weather season and trough production in the winter.

In fact, all four of the major fields considered here (accounting for about 50 per cent of total UKCS output), as well as several smaller fields, are on pipeline systems. The weather should not be a factor influencing their output. Besides, peak production for total UKCS occurred in the winter and trough production in the summer. We may conclude that our second factor does not help much in explaining observed seasonality.

The third factor is maintenance. A close examination of this factor suggests that maintenance may be classified into several types. First, there is unplanned maintenance which must be attended to immediately. By its very nature this should produce random fluctuations in output with no particular seasonal pattern.

Secondly, there is planned maintenance that does not particularly require calm weather. These works can be carried out at virtually any time of the year. They may or may not necessitate a shut-down of production. However, since their timing is not tied to particular months or seasons, they cannot explain a regular yearly cycle in production.

Thirdly, there is planned maintenance that requires calm weather and is therefore carried out in the summer months. In some cases such planned 'summer' maintenance is not associated with a production shut-down or with a partial output adjustment. It may, for example, be carried out on plant or equipment available in duplicates, or on parts of the platform structure that do not interfere with extraction plant. In other instances, however, maintenance involves components of the system that are unique (e.g. main pipeline valve, flare stack, etc.). It is then impossible for operations to continue until repairs are completed so a shut-down is required.

All these comments on maintenance also apply to new works.

Thus the construction of, say, a satellite platform need not always interfere with production from existing facilities; but the tying-in of new pipes to a pipeline system, which is usually planned for the summer, necessitates a shut-down.

In short, maintenance and new works which need to be carried out during the summer because of the need for good weather, and which involve shut-downs or output adjustments, are the only technical factors which, at first sight, may explain a seasonal output pattern.

To sum up; we cannot be entirely certain from the evidence examined so far that technical factors, on their own, account for the whole of the seasonal movement in total UKCS oil production. It is thus important to examine whether the view (sometimes expressed outside the North Sea oil industry) that economic factors influence the output pattern contributes to the explanation.

5.6 UKCS Output Seasonality and Economic Choice

The economic hypothesis states that oil-producing companies may vary the rate of utilisation of existing capacity to suit changes in trading conditions.

Two different formulations may be propounded:

— The first is that companies respond to fluctuations in spot prices by an accommodating change in production. When spot prices fall output is reduced, and when spot prices rise supplies are increased.
— The second hypothesis is that companies take the view that oil demand displays a regular seasonal pattern with a peak in the winter and a trough in the summer. They therefore plan UKCS oil output to match – hence the marked output rises in winter and the significant decline in summer.

The first hypothesis is not consistent with a regular seasonal output pattern save in the case where spot price movements display a similar regular pattern. The evidence is that they do not. For example, spot prices were weak in the winter of 1983 and they rose significantly in the third quarter (summer) of that year. In 1984 spot prices declined fairly sharply in June and July (late spring, early summer). This may be clearly seen in Figure 5.7.

The second hypothesis is consistent with a regular seasonal pattern of output. It relates, however, to a perceived (not

Figure 5.7   Spot Price of Brent. July 1982–December 1985

necessarily to an actual) seasonality in demand, and is therefore difficult to test.

Statistical analysis confirms that spot price movements differ from the output pattern. We also estimated an equation in which production is related, among other things, to spot prices. The estimation was made for each of five major fields, total UKCS production, and total UKCS minus Brent and Forties. In all but two cases the coefficient of the price variable turned out to be statistically insignificant. This particular model, in five out of seven cases, shows no evidence whatsoever of supply sensitivity to prices. The two exceptions are Brent and Dunlin. A regression for Brent was estimated over a short period (January 1984–February 1985); the coefficient of the price variable turned out to be significant and positive, meaning that output moves in the same direction as prices. However, the value of the coefficient is small, implying that a $1 change in the Brent price is associated with an output change of 21,200 b/d. The regression for Dunlin shows strong output sensitivity to the spot price.

Another variant of the model, in which the Brent netback price was used instead of the spot price, was also tried. It can be argued that oil companies are more likely to treat netbacks rather than spot crude prices as indicators of the true state of the market, and that their economic decisions are more strongly influenced by netbacks than by spot prices.

The results generally turned out to be statistically insignificant in all the cases considered but one. The only exception is Forties: the coefficient obtained is statistically significant when the independent variable is the Brent netback series for basic refining at Rotterdam.

To sum up the results on the price/output link, the attempt to establish a statistical relationship between oil output changes and price movements, whether spot prices or netbacks, failed in all cases except three: Brent and Dunlin with spot prices, and Forties with one particular netback series.

An alternative economic hypothesis postulates that companies attempt to plan production in accordance with the expected pattern of oil consumption. This is not the same as saying that output responds to prices, because consumption and price may not move together.

We estimated statistical relationships between UKCS total oil production (also Brent and Forties taken singly) and UK refinery output as a proxy for consumption. It must be stressed that refinery output is an imperfect indicator of consumption because of inventory changes but this is the best variable available. We found a correlation between crude oil production and refinery throughput for UKCS total output and for Forties but not for Brent. Furthermore, the statistical analysis suggests that oil production is characterized by seasonal patterns which are independent of variations in refinery output.

Given these inconclusive results we turn to a different hypothesis, which suggests that the fluctuations in UKCS production are essentially due to Brent, which is both a large and a particularly difficult field. The fluctuations in Brent output could be partly due to the constraints imposed by a high gas/oil ratio. When the volume of gas sales changes because of seasonal demand, oil production is varied accordingly. Gas is sold on contract and oil is largely disposed of on spot. Further the flaring of gas is strictly regulated. These considerations mean that gas sales drive oil production in Brent. We tested this hypothesis by relating Brent oil output to gas consumption in the UK, and the results obtained were statistically significant.

Gas probably explains why Brent oil output is driven flat out in the winter and allowed to drop in the spring or the summer during which maintenance and repairs can be fitted in. Yet the behaviour of the Brent field does not account for the whole seasonality of

UKCS oil production since we found significant output fluctuations in Forties and Dunlin and (with less confidence in the results) in Fulmar.

Our conclusions about the causes of oil output fluctuations in the UKCS are that technical factors are important but do not explain everything. The view that output is also sensitive to economic factors is not well borne out by the evidence but cannot be dismissed altogether. Whatever economic influence may exist on output, the relevant factor seems to be expected oil demand rather than prices. Brent is responsible for a large part, but not all, of the seasonal output fluctuations in UKCS; and Brent fluctuations are largely due to gas. It is intriguing to find that fields operated by Shell and BP (Brent, Forties, Fulmar, Dunlin) are more seasonal than fields operated by smaller companies such as Occidental (Piper, Claymore) and seemingly more responsive to economic factors. One should not read too much into this finding; but simply recall that large oil companies are more likely to take demand patterns into account in making output decisions.

PART II

THE INSTITUTIONAL FRAMEWORK

GENERAL INTRODUCTION TO PART II

An analysis of certain aspects of the institutional framework is of considerable relevance to the study of the market. The various economic agents in a market – producers, buyers, traders, brokers – operate within a system of regulations and procedures which shapes their behaviour. The fiscal regime provides and removes incentives; it also frequently distorts price signals and causes inefficiencies. Institutions such as national oil corporations sometimes play a major role in the market as economic agents engaged in competition with other buyers and sellers; yet they usually have objectives of their own and display specific modes of behaviour.

In Part II we shall describe:

— the licensing and production policies of the UK and Norway (Chapters 6 and 9)
— the place and role of the major national oil corporations, BNOC and Statoil (Chapters 7 and 9)
— the fiscal regime applied to oil in the two countries (Chapters 8 and 10)

Licensing policies have some influence on long-term supply patterns, whether or not they reflect a coherent depletion policy; but their effects on the day-to-day operation of markets tend to be indirect and not very significant.

Production policies can be used for the purpose of short-term market intervention, or for the pursuit of depletion objectives in the longer run. In either case they tend to influence the behaviour of relevant market agents. Neither the UK nor the Norwegian Government so far seems to have used this particular instrument for relieving markets of temporary or structural supply surpluses. However, it is important to identify the powers available to governments in these matters in order to assess the constraints under which they may act in the future, should production controls become desirable or necessary.

National oil corporations have played (and Statoil continues to play) an important role in the development of the North Sea and in the trading of North Sea crudes. BNOC, and to a lesser extent Statoil, have also performed pricing functions with an impact on both the North Sea oil industry and the world petroleum market. The significance of this impact is a question for investigation.

The oil fiscal regimes have a variety of effects on the investment behaviour, the profitability and the trading methods of companies. We shall describe the main features of the fiscal systems because of their inherent interest; but we shall only be concerned with one set of effects, those relating to the trading pattern. In the UK the taxation system has encouraged a phenomenon known as 'tax spinning' on the part of integrated oil companies with production interests in the UKCS. A company is said to spin its oil when it finds it advantageous, for tax reasons, to sell part of its output at arm's length instead of retaining it for direct supplies to its refineries and affiliates.

CHAPTER 6

UK LICENSING AND PRODUCTION POLICIES

6.1 The UK Institutional Framework. A Political Introduction

The UK Continental Shelf has been explored for hydrocarbons and developed by private enterprise (and also earlier on by some national corporations) operating within a fairly elaborate framework of legal requirements, administrative regulations and fiscal impositions.

The institutional framework was more constraining in the period 1975–9 under the Labour Government than under its Conservative successor (1979 to date). In 1979 the Conservatives decided to relax some of the administrative regulations relating to the role of BNOC as adviser to the Government and as its watchdog in operating committees. This decision was implemented in 1982, when the Government privatised the upstream activities of BNOC, and in early 1985 it announced to Parliament its intention to abolish the corporation. By doing so, the Government removed an important part of the institutional structure.

Labour was inclined to regulate the activities of private enterprise in the North Sea, and it promoted the development of a large national oil corporation which put the public sector in control of a portion of the North Sea. The Conservatives seem to want less regulation and no public sector involvement in the oil industry.

Though the Conservative Government chopped up and then abolished BNOC, it did not restrict the existing powers and rights of the Department of Energy. These powers are extensive in some areas and fairly limited (*de facto* rather than *de jure*) in other spheres. The main source of power in the Government's hands is the discretionary granting of petroleum production licences.

On oil taxation both political parties, when in power, have displayed the same appetite for maximizing revenues, and Petroleum Revenue Tax has been progressively increased from 45 per cent at the start to 75 per cent today. The Conservative Government, possibly in the belief that oil prices were set to rise for the forseeable future, went even further and introduced the Supplementary Petroleum Duty (1981 to end 1982). In the event prices started falling and the tax proved highly unpopular with the oil industry. It was replaced by Advance Petroleum Revenue Tax in 1983, which is now being phased out. The Conservatives also reformed Corporation Tax in the 1984 Budget. Both changes will ease the fiscal burden on oil companies.

6.2   The Licensing of Exploration and Production

All rights to the seabed, subsoil and their natural resources are vested by UK legislation in the Crown. The Government is empowered to issue licences for exploration and production of hydrocarbon reserves. Production licences (to be distinguished from exploration licences) entitle the licensee to explore for, develop and produce hydrocarbons in the concession area without limit to the depth of drilling for a period of up to thirty years subject to agreement on plans for development. (There have been nine rounds of UKCS exploration and production licences to date.)

The licensing method generally followed by the UK is the discretionary award. Of course, discretion is guided by a large number of principles, criteria and stated preferences which may vary from round to round according to Goverment objectives and changing circumstances. Nevertheless the discretionary system is a powerful and effective weapon in the hands of the Government, and can be used to ensure that both licensees and would-be licensees co-operate with Government policy. An alternative licensing method, which market economists tend to advocate, is the auction system. The Government experimented on a very small scale with the auction system in the fourth round (1972) by putting to tender fifteen blocks. In the seventh round (1980–81) companies were allowed to indicate their choices from any unlicensed blocks in a defined area of the North Sea; a £5 million premium was payable on the granting of the licence.

The first four rounds were held in 1964, 1965, 1970 and 1971–2. Their aim was to speed up exploration for and development of UKCS oil and gas resources, and accordingly the terms tended to be generous. The interval between rounds was irregular and the licensing strategy, which perhaps reflected the Government's lack of experience of oil affairs in these early days, tended to be haphazard. These rounds involved a gross total of 2655 blocks on offer. The number of licences issued was 245 and they covered 863 blocks. All fields in production in 1985 had licences from these first four rounds.

The next rounds were held in 1976–7, 1978–9, 1980–81, 1982–3 and 1984–5. All these took place after the Petroleum and Submarine Pipelines Act, 1975. In the fifth and sixth rounds, under a Labour Government, BNOC was granted an automatic 51 per cent stake in each block (except where the British Gas Corporation was a participant). However, in the sixth round, BNOC's co-licensees had to carry the share of exploration costs, reimbursable if a discovery was made and declared commercial (strictly speaking, this requirement was not mandatory).

The fifth to ninth rounds differed in other respects from the earlier four rounds. For example, the basis of royalty computations was changed from 12.5 per cent of the well-head value of production (first four rounds) to 12.5 per cent of the landed value of production as defined for PRT purposes.

The fifth to eighth rounds involved a gross total of 381 blocks offered by the Government. The number of licences issued was 199 and the number of blocks covered was 246. The ninth round involved 195 blocks on offer; 147 applications were made by 130 companies and ninety-three blocks were awarded, of which thirteen were auctioned.

6.3   Government Control over Oil Production

The Government influences in a very broad and loose way the long-term oil production pattern through the issue of licences. The flow of blocks to licensees depends on the number of blocks offered in each round, the terms and conditions of licences (which naturally determine the companies' response) and the interval between rounds.

However, the licensing flow is not a very good instrument of depletion policy. The instrument is too blunt and its effects are uncertain. There is no close relationship between the number of blocks offered, the amounts of oil that will eventually become available for production and the future time pattern of output. All depends on the rate of discoveries and the type of development that these discoveries will then elicit.

The Government has a more definite influence on the long-term production profile once commercial discoveries are made and companies seek approval for development and production plans. These plans must be submitted in document known as Annex B. The Government has considerable powers at this stage as it can reject the proposals, defer approval, or grant a staged consent. The criteria applied in examining proposals include the compatibility of the proposals with good reservoir management to ensure optimum recovery, gas flaring regulations and similar technical considerations. The Government makes a broad judgement based on the national interest.

The Government may either give consent to the proposals under Annex B, and attach such conditions as are appropriate, or may give approval to a development and production programme, which is set out in another document, Annex A, but such blanket approval has only been given for Argyll, Montrose and perhaps Fulmar and Cormorant.

All the early fields up to Murchison were authorised on short-term consents (renewed every three to six months). The other fields have mainly been approved using a two-stage approval, a first-stage approval of around four or five years allowing unfettered production, and a second-stage consent for a further period which could enable the Government to require changes in production provided due notice was given and after licensees had had an opportunity to argue their case. Some short-life fields are granted only single-stage approvals or consents for a period of years, again allowing for unfettered production during that time.

The development plan of a field, given the expected size of the hydrocarbon reserves and the characteristics of the reservoir, determines fairly closely the production pattern. Some flexibility in matters of capacity exists *ex ante* in the form of choices between different types of development but this flexibility is reduced and capacity is virtually a given datum after the implementation of the selected development plan.

Through the approval procedure, which involves both the types of development and the implementation schedule, the Government has some influence on the future production profile. Approval also covers an annual production profile, and is usually given with a tolerance of plus or minus 5 per cent. Of course, a company can revise this output estimate in the light of the experience gained through development and production. It would then seek a revision of the agreed profile in the approval.

The only effective method of output control in the short run is one which relates directly and immediately to producing fields. In principle, such an instrument is available to the Government in the form of 'limitation notices' which the Secretary of State for Energy can serve on operators. In law the Government has complete control over the level of oil production. These powers derive from the Petroleum (Production) Act, 1934 and can be exercised in case of national emergency or when necessitated by the national interest.

The latter concept is fairly broad and open to many different interpretations. Except in emergencies, consultation with the companies concerned has to take place before an output limitation is imposed and six months' notice has to be given to all companies. The legal aspects of these procedures are not very clear to us. In practice the Government has regarded these as 'reserve powers' and UKCS companies have been given solemn assurances by successive Governments that they will not be used for the time being.

This is the now famous Varley assurance given in December 1974, which reads as follows:

> I wish, therefore, to assure the oil companies, and the banks to which they will look for finance, that our depletion policy and its implementation will not undermine the basis on which they have made plans and entered into commitments. Our future policy will be based on the following guidelines:
>
> *(a)* No delays will be imposed on the development of finds already made or on any new finds made up to the end of 1975 under existing licences [i.e. First to Fourth Round licences]. If it should prove necessary to delay the development of finds made in 1976 or later, there will be full consultation with the companies so that premature investment is avoided.
>
> *(b)* No cuts will be made in production from finds already made, or from new finds made before the end of 1975 under existing licences, until 1982 at the earliest, or until four years after the start of production, whichever is the later.
>
> *(c)* No cuts will be made in production from any field found after 1975 under an existing licence until 150 per cent of the capital investment in the field has been recovered.
>
> *(d)* If we later need to use these powers we will have full regard to the technical and commercial aspects of the fields in question; and this would generally limit cuts to 20 per cent. at most. We shall be consulting the industry on the period of notice to be given before any reduction in production comes into effect.
>
> *(e)* In deciding on action to postpone development or limit production the Government will also take into account the needs of the offshore supply industry in Scotland, Wales and other parts of the United Kingdom, for a continuing and stable market.[1]

The Varley assurance which expired in the main at the end of 1982 was renewed by Mr Lawson until the end of 1984. On several

[1] Statement by the Secretary of State for Energy on Depletion Policy, 6th December 1974, reprinted in *Brown Book*, 1975.

recent occasions the Government has made it clear that it has no intention of imposing production limitations on firms operating in the UKCS.

6.4   Conclusion

To conclude, the Government can limit oil production in the North Sea if it determines that the national interest calls for such measures. The relevant powers are clearly specified by the law of the land. Further, the Government has considerable reserve powers. Ministerial assurances are not legally binding, particularly in cases where a national emergency or the national interest is involved. The contractual obligation that the Government may have incurred *vis-à-vis* companies to give notice and/or consult before serving limitation orders is not an insuperable constraint. The recent Government actions leading to the abolition of BNOC show how quickly and expediently the Government can persuade companies to relinquish contractual rights (e.g. profitable participation agreements) without litigation and without long delays.

However, the production policy, so far, has been to allow firms to deplete oil at a rate consistent with the maximization of their private economic interests so long as best-practice technology and methods of extraction are used, associated gas is properly disposed of or conserved, and the recovery factor of the reservoir is optimized throughout the life of the field.

The main conclusion is that the Government so far has not attempted to use production policy either for long-term depletion objectives, or for short-term intervention on the world petroleum market in support of pricing objectives. Different types of intervention have taken place on occasion, such as attempts to persuade oil companies to increase their purchases from, or reduce supplies to, the spot market, or to refrain from requesting BNOC to change the term price. These actions may have an immediate, though very short-lived, psychological effect but do not influence long-term trends.

CHAPTER 7

THE UK. THE PLACE AND ROLE OF BNOC IN THE INSTITUTIONAL FRAMEWORK

7.1 Introduction

BNOC was a most important component of the institutional system within which the market for North Sea crudes operated from 1975 to 1985. The Corporation was a very large primary seller of UKCS oil; as such it was bound to influence the formation of price expectations as well as actual prices in both the spot and term markets. In addition to this trading role which involved indirect price effects, BNOC had a more direct pricing function. The Corporation had to determine (through negotiations) a term price for its purchases of participation crudes and for its contract sales to third-party customers. This BNOC term price, which had some of the characteristics of an administered price, and yet had to be determined with reference to actual market conditions, became an important pricing parameter for petroleum in international trade.

7.2 Origins and Initial Objectives

BNOC was established in 1975 by the Petroleum and Submarine Pipeline Act. The apparent motive behind its formation was to shelter the UK from the consequences of an oil supply shortfall in the event of a major international crisis similar to the 1973 embargo and production cut-backs.

In 1973–4, the Conservative Government failed to persuade British multinational oil companies to divert oil supplies to the UK. The companies argued that they were bound to treat all their customers in the same manner and followed a pro-rationing policy which shared the production shortfall more or less proportionately all round. There is no doubt that the British Government was unhappy about this situation. When UKCS oil began to come on stream (1975) the Government naturally sought to establish a system that would provide control over the disposal of UKCS production in the event of an emergency.

Whether the objective of controlling supplies in an emergency (the so-called strategic argument) is best served by the establishment of a national oil corporation is a moot point. It is interesting to note that this strategic argument has been consistently put forward by both Conservative and Labour Governments. It was used to justify the continuation of BNOC as a trading agency after the privatisation of Britoil in 1982. More recently, the argument was mentioned again as one of the reasons for retaining the arrangements under which royalty oil is paid in kind to an official agency after the abolition of BNOC.

Many analysts believe, however, that the strategic argument is not very convincing in the sense that it does not require the establishment of a national oil corporation. The Government can take immediate and full control over the disposal of UKCS oil production in an emergency. This can be done either by means of an emergency statute or by invoking existing legislation which already gives the state wide-ranging, if not absolute, powers when national security or the national interest is at stake.

The power to retain all UKCS oil in the country may be limited by international commitments entered upon by the British Government with the EEC and the IEA; but these commitments would be binding whether oil supplies were controlled by a national corporation or by some emergency statute.

We should also add that the problems encountered by the UK Government in 1973 in its attempt to persuade oil companies to divert supplies to the UK would not have been entirely removed in the 1970s whichever policy was followed. Neither the establishment of BNOC nor an emergency statute would give the British Government control over the disposal of non-UK oil by multinational oil companies. In the period during which UKCS production was lower than domestic consumption, the critical factor in emergencies was access to non-UK oil, and this access would only be secured through credible international agreements implemented in a crisis by all parties.

The case for establishing BNOC would seem to relate to two other arguments:

— the desirability of having a public sector interest in an important industrial/natural resource sector
— the need for the Government to obtain information and advice on matters of oil from a company with industrial experience and with a seat on the Government side of the negotiating table

The first argument is favoured by one political party and totally rejected by the other. In the UK fuels other than oil (coal, gas and nuclear) are in public sector hands. There may, therefore, be a non-ideological case based on consistency and expediency for the Government to retain a limited but not insignificant public sector interest in petroleum.

The second argument is sometimes dismissed on the grounds that a Government that leaves private companies with a proven industrial and commercial record to operate freely within a firm but simple policy framework does not need much outside advice. Another criticism is that a national corporation tends to develop interests and objectives of its own and often finds itself in conflict with Government rather than on its side. Whatever the validity of these counter-arguments it remains true that: (a) information and knowledge are essential to good decision making, and (b) information and knowledge on petroleum are not available with sufficient reliability outside the oil industry. The government of a major oil-exporting country needs to draw on direct industrial experience, and may find the access channel provided by a national corporation valuable in certain circumstances.

For a short period (1976–9) BNOC had some influence – which some consider to have been useful and others to have been negative – on the exploration, development and production in the North Sea:

— First, BNOC acquired some equity interests in a number of fields, either through the provisions for equity participation laid down in the fifth and sixth rounds, or through direct awards made by the Government to BNOC outside regular rounds (eleven blocks given under exclusive licences in April 1978).
— Secondly, BNOC had the right to be represented and to vote on all operating committees. This gave the Corporation access to information about all fields under development or in production.

Not surprisingly the private oil companies resented and disliked the interference of this official watchdog. The most serious charge levelled against BNOC by the oil companies was that its enforced association with them through the operating committees was incompatible with its role as a competitor for prospective exploration and production licences, and that this had led to a decline in exploration and development activity in 1978 with a consequent delay in the attainment of self-sufficiency.

These episodes of BNOC's existence were short-lived. BNOC's powers were curtailed by measures announced in July 1979 by the new Conservative Government. In 1982 the upstream interests of BNOC were privatised through the formation of Britoil.

Despite all this BNOC remained an important part of the North Sea institutional framework for ten years. It played a role because it was a major seller of oil obtained in royalty and participation crude. BNOC was given access to oil because of the strategic objective, and this went some way towards the achievement of the ideological objective. The latter, involving public ownership of productive assets, called for a substantial equity stake in UKCS fields; but such an undertaking would have proved to be very expensive. A cheap and simple way to achieve these objectives, albeit imperfectly, was to give BNOC participation rights in crude oil.

The participation arrangements, together with the disposal of royalty oil on behalf of the Government, had two major implications:

— First, BNOC became a very large trader.
— Secondly, BNOC acquired a pricing role since it had to define a price for its participation purchases and its subsequent disposals to term customers.

7.3   Trading and Pricing Roles

(a) Trading. As a trader BNOC had to dispose of oil acquired in the form of: (a) royalties in kind, (b) participation less sales-back to six integrated oil companies with refineries in the UK (Amoco, BP, Esso, Mobil, Shell and Texaco) and (c) assignments from small licensees who found it convenient to dispose of oil in that way. In early 1985 the amount at BNOC's disposal was approximately 800,000 b/d. This made BNOC the largest primary seller of North Sea crudes in the world, with availabilities twice as large as those of any of the big three majors – BP, Shell and Esso. The amounts at BNOC's disposal are shown in Table 7.1.

BNOC disposed of this oil in different ways:

— sales-back of royalty oil under long-term contracts to producing companies with a buy-back agreement
— sales-back of participation crude under special arrangements to integrated companies without an ordinary buy-back agreement
— exchanges and swaps
— sales under term contracts
— direct sales on the spot market

Table 7.1:   Crude Oil Volumes at BNOC's Disposal. 1978–84. Thousand Barrels per Day

Source	1978	1979	1980	1981	1982	1983	1984
Equity	11	64	85	105	78	–	–
BGC equity	13	18	19	17	16	–	–
Purchases	24	49	42	71	115	178	175
Royalty in kind	–	28	180	217	229	246	260
Participation (net)	51	322	235	372	365	403	435
Net Sales	99	481	561	782	803	827	870

Note: These figures differ from BNOC's primary availabilities shown in Chapter 3 because they include secondary transactions and treat part of participation crude as purchases.

Source: BNOC, *Annual Report*, 1984

In some instances BNOC purchased additional oil for onward sales but these operations were dictated simply by the need to correct imbalances arising between sales commitments and availabilities.

Until mid-1984 BNOC was generally able to dispose of most of its oil under term contracts. Sales on the spot market were not very significant. Recourse to the spot market took place whenever availabilities exceeded commitments under term contracts, either because of unanticipated increases in accruals or because of reductions in the amounts covered by contracts through phasing-out of purchases by customers.

In the period to 1984 spot prices were not always lower than the BNOC term prices. There were episodes during which spot prices were higher, making BNOC's recourse to the spot market, occasional as it may have been, clearly profitable. The situation changed in the middle of 1984 when spot prices began to remain consistently (with insignificant exceptions) below the BNOC term level.

As this situation persisted BNOC's contract customers began to withdraw, and the volumes that the Corporation had to sell either directly on the spot market or through contracts with flexible (spot-related) price clauses increased. This involved trading losses. The most important consequence, however, was the emergence of a very large seller putting increasingly large quantities on a weak spot market.

The peculiarity of BNOC as a trader was its inability to determine freely, or at least influence, the volume of its supplies. At any point in time, the volume of BNOC's supplies was a given proportion of output, an independent variable lying completely outside BNOC's discretion.

Any normal trader continually makes two autonomous decisions: (a) he freely determines the amounts he wishes to purchase and seeks to obtain them either on the spot market or through term contracts; (b) he sells amounts at his disposal, drawing or adding to inventories as appropriate, on the best terms he can (expect to) obtain.

In contrast, BNOC had no discretion as to the amounts acquired and was obliged to dispose of these quantities almost immediately because of a total lack of storage facilities.

An agency, as constrained in terms of the volume of acquisitions and the volume of sales as BNOC was, cannot really be likened to a trader without distorting the meaning of this concept. Such an agency can respond to changing market conditions only by changing its prices, because it has virtually no scope for regulating the quantities at its disposal. It also suffers from a clear disadvantage when competing with other traders since its ability to manoeuvre involves one degree of freedom (price) instead of two (volume and price).

A legitimate question to ask is whether this phenomenon had an effect on spot price movements or whether it was price neutral. A fairly simplistic answer, which many observers are too ready to make, is to say that the price effect was neutral because the amounts BNOC put onto the spot market were equal to the amounts BNOC term customers wished to purchase from that market. This answer misses however some aspects of the situation. First, BNOC had little room for manoeuvre on the spot because of its need to dispose of its availabilities immediately while buyers had some control over the timing of their purchases. Buyers enjoyed flexibility through recourse to inventories and access to diverse sources of supply. Secondly, the emergence on the spot market of a large agency perceived by many as a distress seller had an adverse effect on expectations; and there is no doubt that expectations play some role in the process of price formation. Thirdly, additional supplies on the spot market cause a temporary imbalance (in the form of excess supplies) and a reduction in prices. The correction of the imbalance through a later increase in purchases need not bring the spot price back to its initial level. This is because a fall in spot prices can bring about a reduction in official prices outside the North Sea (as happened after the crisis of October 1984–January 1985), and the new lower official prices will tend to influence the subsequent spot price movements.

(b) Pricing Role. The pricing role of BNOC was naturally related to its trading activities. BNOC was a price administrator in a very particular sense. It posted a price for the purchase of participation crude which was supposed to reflect the exchange value of a barrel of oil for willing buyers and sellers striking a bargain in a free market. In ideal conditions the BNOC administered price is nothing but a market price.

In the complex conditions of the real world it is difficult to identify this 'market price' with the true outcome of a bargain between willing parties. Those who sold oil to BNOC and those who bought oil from BNOC constituted two heterogeneous groups. The sellers did not have identical circumstances and objectives. The buyers were engaged in a bargain with BNOC, not with those who initially supplied the Corporation with participation and royalty crude.

First consider the sellers, or more precisely the producing companies that were legally or contractually bound to transfer part of their output to BNOC. Producing companies (oil or non-oil) that were net sellers of oil always sought the highest possible price. Producing companies that were net buyers of oil had an economic interest in lower prices. The willingness of these two subgroups of suppliers to accept a price proposed by BNOC depended on different sets of factors. Further, the willingness to accept or reject a price set by BNOC is different in nature from the concept of a market bargain between willing buyers and sellers. In the former case the seller is not entirely a free agent since he is obliged to supply participation oil. He does not have the option of seeking alternative buyers, of 'shopping around' for the best deal. In case of disagreement with BNOC he can have recourse to arbitration which is different – more cumbersome and less informative – from the normal market process of 'shopping around'.

Consider now the buyers of BNOC oil. Their bargaining power *vis-à-vis* BNOC was stronger than the suppliers' because they could: (a) simply refuse to enter into a purchasing contract if they did not like the proposed price, or (b) invoke phasing-out clauses if they had already entered into a contract and come to disagree with the ruling or a newly proposed price.

Put differently, there was an asymmetry between the relationships of BNOC with its suppliers and with its buyers, since the balance of power in the two cases was different. The subtle difference in these relationships and the fact that in reality BNOC was an intermediary between suppliers constrained by participation agreements and buyers free to enter into and to get out of contracts together vitiate the concept of the pure market-bargain price.

BNOC's intermediation made it possible to administer prices with varying degrees of closeness to market conditions. However BNOC's freedom in this matter could only be exercised within narrow limits. To set administered prices above market levels (as reflected, say, by the spot market) would lead to trading losses causing embarrassing requests to Parliament for additional appropriations and political hostility. To set them below market levels would lead to litigation with dissatisfied producing companies.

The pricing decision was made more complicated by another set of choices. BNOC was also faced with the following options:

— either to vary its term price in fairly close harmony with the market in order to take full benefit of a tight supply situation (as in 1979–80) and to avoid losses in a slack market (as in late 1982, early 1983 and October 1984)
— or to keep its term price in harmony with the OPEC price structure, forgoing potential benefits in a tight market (as could have happened but did not occur in 1979–80) and incurring losses in a slack market (as did happen during certain months in 1984 and in the first months of 1985).

In 1979 BNOC followed the market with rapid and significant upward adjustments to its term prices. In fact BNOC's responses were so swift as to precede official price rises by OPEC countries. BNOC was following the market and in the process bringing additional revenues to the UK as well as making trading profits. The criticisms voiced by some OPEC 'moderate' members such as Saudi Arabia, which argued that the UK was contributing to an unwelcome and dangerous price explosion, were consistently dismissed on the grounds that the UK was a very small producer with no influence on market trends. Indeed UK oil output and exports were small at the time.

In 1983, BNOC, most probably on instructions from the Government, followed OPEC. OPEC had then become concerned about the effects of changes in the BNOC term price on its own official price structure. From March 1983 until October 1984 BNOC (or more precisely the UK Government) was careful to fix the term price at a level consistent with the OPEC price structure.

One objective of this policy was to avoid the accusation of causing a disturbance on the world petroleum market through a reduction in the BNOC term price. The more fundamental objective, perhaps, was to avoid an oil price collapse which might damage UK economic interests.

The policy of administering the BNOC term price in line with the OPEC price structure became unworkable in mid-1984. The gap between spot and term prices caused trading losses requiring supplementary expenditure appropriations from Parliament. These trading losses encouraged various UK circles – particularly the Press, Tory MPs and probably some members of the Government – who held the view that BNOC was serving no useful purpose, to step up their campaign for the abolition of BNOC. In March 1985, the Government gave in and announced its intention of abolishing the Corporation.

7.4 Conclusions

BNOC was created for strategic and ideological reasons. It never fulfilled a strategic role (not even in 1979 during the Iranian crisis) during its lifetime. The ideological motive did not serve it well; although it contributed to its establishment, it also led to its abolition when a Government with a different ideology came to power.

BNOC instead performed trading and pricing roles. The main difficulty faced by the Corporation in the performance of these roles was the need to reconcile a number of conflicting objectives. The Corporation was asked:

— to retain a term price related to the OPEC price structure in a soft market, and
— to avoid politically embarrassing trading losses.

At the same time, BNOC was denied the means of becoming an efficient trader (the necessary discretion over the acquisition of oil and the availability of storage facilities); and the UK Government was denying itself the means of reconciling the objectives through production controls and supply intervention.

In these circumstances, the failure to fulfil conflicting objectives was inevitable, and this led to the abolition of BNOC. This was an easy way out of a difficult dilemma. The alternative solution, which a Conservative Government could not contemplate, was a restructuring of BNOC to transform it into a genuine national oil corporation.

While in existence BNOC played a part in contributing to the stability of oil prices between March 1983 and mid-1984. Considering the large imbalances between supply and demand on the world petroleum market, this was no mean achievement.

It is often forgotten that though spot price movements ultimately have an impact on term prices and force changes in their levels, term prices also exercise an influence on spot prices. Of course, this influence is only fully effective when backed by output restraints. In a soft oil market, exporting countries' governments need all the instruments available for defending price levels. A term price structure is one of these. To abolish it is to deprive oneself of a useful policy instrument.

There were, however, many drawbacks relating to BNOC's trading and pricing roles. It is difficult to draw up a balance sheet of the positive and negative contributions of BNOC to the North Sea market for crude oil, and to the stability of the world petroleum market. In the North Sea, BNOC appeared to be a distress seller on certain occasions. In the world petroleum market, BNOC triggered (but did not cause) three major price crises – in early 1982, early 1983 and October 1984.

All things considered, we can only state that the history of BNOC does not provide a conclusive case either for or against the establishment of a national oil corporation in the UK. The initial concept of BNOC was too ambitious and proved to be counterproductive. The privatisation of upstream interests (Britoil) left a trading rump which was not viable. The subsequent abolition was motivated by immediate considerations of political expediency. These successive failures of conception point to a more fundamental problem: the ambivalence of official UK views on the critical issue of the price of oil, and on the role that the UK should play on the world petroleum market.

CHAPTER 8

THE UK. THE OIL FISCAL REGIME

8.1 Introduction

Fiscal regimes are rarely neutral; they have an impact on the behaviour of economic agents affected by the regime, and through them on the functioning and efficiency of markets. The purposes of this chapter are threefold:

— To describe the main features of the UK oil (and gas) fiscal regime (sections 8.2 to 8.5).
— To present some data on effective rates of oil taxation for UKCS producing oilfields (section 8.6).
— To analyse the impact of the UK regime on the market behaviour of oil companies. For the purposes of this study the relevant fiscal impact on the market is the phenomenon already referred to, known as 'tax spinning' (sections 8.7 to 8.9).

It is useful, however, to introduce the description of the different taxes by a mention of two important features of the oil fiscal regime. The first refers to pricing principles and the second to the concept of a fiscal ring-fence.

(a) Pricing. The UK oil fiscal system bases the valuation of crude oil produced by an oil company on market prices, that is on the realized price of genuine arm's length transactions. An alternative valuation method adopted by some countries involves the use of administered instead of market prices. The 'norm price' system in Norway (see section 10.7 below) is an example of this alternative method.

It should be emphasised from the outset that the UK legislation adopts unambiguously the 'market price' valuation principle for tax purposes, because there seems to be some confusion on this issue among international oil commentators and experts who are not very familiar with the intricacies of the British fiscal regime. The confusion arises because of a loose usage of the term 'tax-reference price' in discussions and commentaries, and because the BNOC price, which was regarded for a while as a good indicator of the market price by the UK tax authorities, is itself construed by many as an administered price. In fact, the term 'tax-reference price' applies indifferently to any price, be it a market or an administered price, taken as the basis for tax valuations. Furthermore, the BNOC price was only considered relevant for tax purposes when the authorities believed that most market transactions were made at that price.

(b) Ring-fence. The UK oil fiscal regime does not in general allow companies to set losses made in other fields and other activities against taxable profits made in any given field for the purposes of PRT. This is known as the field-by-field principle or as the fiscal ring-fence. For the purposes of Corporation Tax the ring-fence is not drawn around each field but it separates the upstream activities of the company in the UK from its downstream activity in the UK and from its other businesses in the rest of the world.

8.2   Main Components of the UK Oil Fiscal Regime

The current fiscal regime for UKCS oil (and gas) production operates in three stages:

— The first stage is the payment of a *Royalty* based on gross field revenues. Royalty is paid in cash or kind at the Government's option.
— The second stage is the *Petroleum Revenue Tax (PRT)* based on individual field profits.
— The final stage is the *Corporation Tax (CT)* based on the total profits of companies resident in the UK.

In addition, the Government introduced a tax known as the Supplementary Petroleum Duty in 1981 which was akin to a windfall profits tax, but this tax was abolished in 1982. Another temporary fiscal measure, the Advance Petroleum Revenue Tax (APRT) which has applied since 1st January 1983 is also being phased out and will be entirely abolished at the end of 1986. APRT, as its name suggests, is designed to bring forward payments of future PRT liability.

We shall briefly describe these various charges and taxes in the following sections. Although most provisions of the fiscal regime apply to both oil and gas, we shall only refer explicitly to oil because gas is outside the scope of this study.

8.3 Royalties

The legal provisions for the payment of royalties to the Government have their origin in the Petroleum (Production) Act, 1934, which was extended off shore by the Continental Shelf Act, 1964. These acts vested the property of all oil in place in the Crown and provided for the payment of royalties by licensees to the Government in its capacity as proprietor.

Royalties are treated by some economists as a tax and by others as a 'rent' levied by the owner. The validity of these different interpretations depends on the context of the analysis. In tax law royalties are treated as an expense to be deducted from income before tax liability is assessed.

The manner in which royalties are calculated for UKCS oil production depends on the round in which the licence was issued. The position is as follows:

— Licences issued in one of the first four rounds: the royalty is 12.5 per cent of the well-head value of production.
— Licences issued in the fifth and subsequent rounds (i.e. since 1976): the royalty is 12.5 per cent of the landed value of production.

There is, however, an exception: royalties are no longer payable on certain fields defined as 'relevant new fields'. Broadly speaking, these are offshore fields in areas other than the Southern Basin which were granted consent for development on or after 1st April 1982.

The difference between the 'well-head' and the 'landed' values of production is in allowances made for conveying, treating and initial storage. Where royalties are calculated on the basis of the well-head value, deductions are made from gross revenues to allow for these costs. This means that the actual royalty rate is less than the notional 12.5 per cent when the basis of calculation is the well-head value. All fields in production in 1984 had their licences issued in the first four rounds (except for a very small amount of Statfjord production); thus the amounts of royalties levied at present (in value terms) represent less than 12.5 per cent of the value of petroleum produced.

In effect royalties amount to 12.5 per cent of (Gross Revenues minus Allowable Costs). Gross revenues during chargeable periods are quantities disposed of multiplied by realized prices plus quantities appropriated by the licensee (say for refining) multiplied by a market price. The principles applied are those used for PRT. Costs that may be set against gross revenues for the royalty calculation on the basis of well-head value are as follows:

— The costs of conveying petroleum from the well-head to the point at which petroleum is valued in the United Kingdom or, in the case of direct deliveries by tankers, to a port in a defined area of Western Europe (49–61°N, 0–13°E).

— The costs of initial treatment, including costs of the processes required to allow petroleum to be safely stored, transported by pipeline, loaded into a tanker or accepted by a refinery.

— The costs of initial storage, that is the storage of a volume not exceeding ten times the planned or actual daily production rate whichever is the greater.

— In most cases 70 per cent of the capital costs of production platforms are admissible for depreciation purposes over eight years or over the life of the field whichever is the shorter.

— Platform operating costs are allowable in full in cases where more than 92.5 per cent of the platform area is used for conveying and treatment. In most other cases the allowance is set at 60 per cent.

— A notional interest allowance is given on the written down value of fixed assets for royalty purposes; working capital is excluded.

The allowable costs also include leasing and hiring, insurance of fixed assets, overheads (both direct and indirect), costs of operation, insurance against pollution, etc. These costs are allowable in full if wholly attributable to conveying and treatment: otherwise they are apportioned according to certain conventions.

Royalties can be paid either in cash or in kind at the discretion of the Secretary of State for Energy. If paid in cash, payments are made for half-yearly chargeable periods (ending in June and December) within two months of the end of the period. If paid in kind, deliveries are made during the relevant chargeable period.

Royalty in kind is required to be delivered at a rate of as near 12.5 per cent of production as circumstances permit. Licensees receive payments on account of the costs of delivery and treatment of royalty oil (effectively equivalent to the conveying and treating costs allowable against royalty in cash) throughout the chargeable period. Adjustments arising from any over- or under-delivery of royalty oil, and/or under- or over-payment of delivery and treatment costs on account, are made two months after the end of the chargeable period. Clearly the taking of royalty in kind rather than cash benefits the Government's cash flow.

For most UKCS oilfields, royalties are paid in kind. In 1984, the exceptions were Argyll, Auk, Beatrice, Buchan, Heather, Maureen, Montrose and Tartan. The Secretary of State can request payment in kind for any field paying royalties in cash by serving a six-month notice, though he must consult the licensees involved before serving such a notice. It is understood that in late 1984 and 1985 some licensees were asked to switch from payments in kind to payments in cash in order to reduce BNOC's availabilities. This reverses the earlier tendency which favoured payment in kind.

8.4 The Petroleum Revenue Tax (PRT)

This tax has its origin in a scheme proposed by the Oil Taxation Bill (November 1974), later adopted with modification and detailed provisions by the Oil Taxation Act 1975 (effective from November 1974). The tax was introduced in response to criticisms made by the Public Accounts Committee of the House of Commons in February 1973. The concern was that the level of Government take would be too low under the existing arrangements.

The Petroleum Revenue Tax has several characteristics that distinguish it from certain standard concepts such as output or profit taxes. Though it bears some resemblance to a unit (per barrel) tax it is in effect closer to a profit tax. However, the rules for calculating revenues and allowable expenditure have unfamiliar characteristics. Furthermore, the introduction of a field-by-field basis (fiscal ring-fence around UKCS fields) for the purposes of PRT gives this tax a significance of its own.

The main features of PRT which distinguish it from Corporation Tax are:

— The 'field-by-field' basis, which means that a company will have as many PRT assessments as it has shares in fields and that company losses in one field cannot be set against profits in others.
— The absence of the usual fiscal distinction between capital and current expenditure.
— The special rules governing the definition of revenue and expenditure which lead to a concept of assessable profits different from the balance sheet concept arrived at by normal accounting procedures.
— The treatment of expenditures which become allowable when the claim for them to be allowed is accepted irrespective of the accounting period in which they are incurred.

— The fact that PRT is a 'prior charge' over Corporation Tax, that is deductible from profits subject to Corporation Tax.

PRT is assessed on the basis of individual field profits. What constitutes an oilfield for tax purposes depends on a determination by the Secretary of State for Energy. There is no appeal procedure once the determination is made but licensees are consulted prior to the decision. It is understood that the determination is made solely on the basis of geological criteria.

The assessable profit for PRT purposes is the total value of oil produced *less* royalties, allowable operating and capital expenditure (the latter being increased by an uplift) and losses from the field that have been carried forward. Since 1983 exploration expenditure incurred elsewhere in the UKCS outside existing fields has been allowable; and in 1981/2 deductions also included payments of the Supplementary Petroleum Duty. Another deduction is the oil allowance – a tax-free volume of oil production designed to help smaller fields.

The concepts used in defining assessable profits require some words of explanation.

Value of oil produced. This is known as the gross profit. It is found by adding:

— Sales of crude produced during the chargeable period and made at arm's length. The relevant price for arm's length sales is the actual contract price. A sale is treated as arm's length if price is the sole consideration and if the seller and the buyer are not connected with each other (that is, are not under common control). In other words, sales between parent and subsidiary or between sister affiliates of a corporate group are not arm's length deals.
— Sales of crude not at arm's length (including crude that is directly appropriated by the licensee for refining purposes). In this troublesome case the relevant price for the valuation of crude is the market price, defined as the price that could have been obtained in a contemporary and comparable transaction with a willing buyer at arm's length (see section 8.7).
— Half the difference between the value of oil not disposed of (or appropriated) at the end of the period and the oil not disposed of at the end of the preceding period. This volume is determined at market prices in the same way as above.

The implications for market behaviour of valuation methods applied to sales not at arm's length, that is to sales between companies of the same corporate group, and to direct appropriation, are analysed in sections 8.7 to 8.10.

Royalties. These payments are deducted if they are made in cash, and the method used for calculating the amount allowed is designed to give relief on an accrual basis. Royalties in kind are excluded by definition from the valuation of gross profits which only include sales, appropriation and inventory charges and do not give rise to any further deductions. Periodical payments (other than royalties) made to the Secretary of State in respect of the licence are also deductible.

Expenditure and Uplift. Expenditures are allowed if they fall under certain prescribed headings. Broadly speaking, these cover all expenditure incurred in finding and producing oil from the field in question. They include preliminary geological surveys, exploration, appraisal, development, operating costs, initial treatment and storage and transport from the field to a delivery point in the UK. Costs incurred in exploration and development including such items as platforms, pipelines, gas separation plants, storage tanks, etc. are inflated by an amount known as uplift in lieu of interest relief on capital expenditure. Until 1979 this supplement was 75 per cent; it was then reduced to 35 per cent. Uplift ceases to be given when the cumulative revenues of a field start to exceed cumulative outgoings.

Exploration and Appraisal Expenditure in Other Fields. The ring-fence round the field has been partially lifted by the Finance Act, 1983, for expenditures incurred after 15th March 1983 in searching for oil in the UK or in appraising a find. (Prior to that date only abortive exploration expenditure was allowable against the PRT payable on an earlier field.) However, expenditures related to fields for which a development decision has previously been made do not qualify for relief against other fields.

Oil Allowance. This is a deduction from gross profits given for each field at the rate of 250,000 tonnes for each chargeable period up to a limit of 5 million tonnes. Since the length of the chargeable period is six months the allowance is given over a period of ten years or over the productive life of the field, whichever is shorter. New relevant fields (fields without development consent on or after 1st

April 1982 and lying off shore outside the Southern Basin) have an oil allowance of 500,000 tonnes per period up to a cumulative maximum of 10 million tonnes. One purpose of the oil allowance is to give additional relief to marginal fields (the allowance, being a fixed quantity, is naturally greater in percentage terms the smaller the output of a field).

Safeguard. PRT is subject to a safeguard provision which in certain periods may eliminate the PRT charge. The purpose of this provision is to prevent PRT becoming a disincentive to the development of small fields. The rule may be phrased as follows: 'Total PRT payable by a licensee in respect of an oilfield for a chargeable period is to be limited to 80 per cent of the amount by which its adjusted profit for that period exceeds 15 per cent of its accumulated expenditure.' The adjusted profit is calculated by adding back to the assessable profit upliftable field expenditure and certain extra-field expenditure allowed during the period. Accumulated capital expenditure is the aggregate of expenditure qualifying for uplift but without the related uplift.

The PRT rate on assessable profits is currently 75 per cent. This rate applies to chargeable periods ending after 31st December 1982. Previous rates were 45 per cent (periods up to the end of 1978), 60 per cent (periods up to the end of 1979) and 70 per cent (periods up to the end of 1982).

PRT is collected on the basis of returns and assessments made for each chargeable period, that is twice a year. The returns must be filed within two months from the end of the chargeable period, and the Petroleum Revenue Tax Act, 1980, requires a payment on account of the liability to be made with the return. This payment corresponds fairly closely to the liability assessed by the tax payer, that is to the liability he would incur if the Revenue were to accept all claims and computations made on the return. The difference between the payment on account and the fiscal liability is made after the Revenue has issued its notice of assessment.

Since 1980 the Government has tried different schemes for accelerating the payment of PRT or for collecting the tax in advance. The procedures for payment on account of the Petroleum Revenue Tax Act, 1980, and the Finance Act, 1980, instituted a system of advance payments.

In 1981 the Government introduced the Supplementary Petroleum Duty (SPD) with the dual purpose of collecting more revenues and accelerating the payment of oil tax liabilities. SPD was charged at the rate of 20 per cent of gross revenue minus the licensee's share of an allowance of 1 million tonnes per year. SPD was deductible from the assessable profits for PRT and Corporation Tax but, as it was payable in monthly instalments, it enabled the Government to collect in advance part of what would otherwise have been a PRT or a Corporation Tax liability payable at a later stage. SPD ceased to apply at the end of 1982.

SPD was succeeded by Advance Petroleum Revenue Tax (APRT). While SPD was partly an additional tax and partly a way to improve the Treasury cash flow, APRT is essentially a method for collecting the PRT liability in advance. Like SPD, APRT is based on the gross profits as computed for PRT with two important adjustments:

— There is an allowance of 1 million tonnes per field per annum.
— A grossing up is made for royalties paid in kind.

APRT rates are 20 per cent (first half 1983), 15 per cent (mid-1983 to end 1984), 10 per cent (to end 1985) and 5 per cent (to end 1986). APRT ceases to apply on 1st January 1987. APRT payments are made in monthly instalments and are treated as a credit set against PRT liabilities. Unused APRT credit is repayable after five years or when the field finally ceases to produce, whichever is the earlier.

There is no doubt that PRT is the most important tax on oil production in the UK. Together with royalties it is the main vehicle through which the Government appropriates a share of the oil rent. The PRT/APRT system is complex because the Government has attempted to achieve a variety of objectives through these measures. Essentially the aims are to obtain the highest possible revenue with the minimum possible delays and in a manner which preserves the incentive to develop both major and marginal fields. The complexity of this tax naturally means a lack of transparency. It has been argued rather convincingly by the Institute for Fiscal Studies that this tax system is inefficient and that, an alternative cash-flow tax is to be preferred.[1] A fundamental taxation reform, however, is unlikely in the immediate future.

8.5 Corporation Tax

Companies are liable to Corporation Tax and oil companies

---

[1] Institute for Fiscal Studies, Committee on the Taxation of North Sea Oil, Report (under the chairmanship of Sir Antony Part), December 1981.

operating in the UKCS are no exception. However, the latter are subject to special provisions, the most important being:

— Corporation Tax for oil companies is levied on the company and not on the individual oilfield. Unlike PRT there is no ring-fence around individual oilfields as for PRT but there is a ring-fence around UK upstream oil operations. This means that an oil company can offset losses in one field against profit made in another inside the fence but it cannot offset losses made elsewhere in the UK or in the rest of the world against profits from UK oil production.
— Relief for unused trading losses brought forward from 31st December 1972 is restricted for sums in excess of £50 million.
— Special rules exist for treating intra-company transfer prices.

The Corporation Tax rate was reduced by the Finance Act, 1984, from 52 per cent (1982/3) to 35 per cent (1986/7). The reduction of the CT rate is to be gradual over that period. At the same time, the first year allowances at which capital costs can be written off for plant and machinery are being phased out, from 100 per cent (1983/4) to zero (1986/7). The amounts not written off under the first year allowance are placed in a pool which is then written off on a 25 per cent declining balance. Development drilling costs and exploration costs are still eligible for 100 per cent first year write-off.

The standard tax rule is that write-offs from a declining pool can only be taken once the asset has been brought into use. This was removed in the 1985 Finance Act, so that the written down allowances can be claimed as soon as the money has been spent.

8.6   Estimated Effective Taxation Rates on UKCS Oilfields

The effective rate of taxation differs from the rate stipulated in law because the former is an average rate arrived at after taking into account all the relevant allowances, tax credits, etc., while the latter is the marginal rate applicable to the last unit of *net* profit or income.

In the UK the law stipulates a 12.5 per cent royalty and a 75 per cent PRT rate, but effective rates for all producing oilfields are lower. The various cost allowances which reduce the size of taxable profits explain the differences (large in certain cases) between marginal and average rates.

The data presented in Tables 8.1 to 8.3 are derived from a model constructed in Aberdeen University by Professor Kemp and Mr Rose. We are very grateful to them for supplying the relevant results of their model.

Table 8.1 shows that in 1984 effective royalty rates (as percentages of gross revenues) ranged between 7.4 per cent for Tartan and 11.4 per cent for Forties. In 1984, the average effective royalty rate for all UKCS production was almost exactly 10 per cent.

Table 8.1:   UK. Effective Royalty Rates as Percentages of Gross Revenues. By Oilfield. 1984.

Field	Royalty %	Field	Royalty %
Argyll	8.8	Heather	8.5
Auk	8.7	Hutton	9.9
Beatrice	8.0	NW Hutton	9.5
Beryl	9.9	Magnus	10.3
S Brae	8.9	Maureen	9.1
Brent	9.3	Montrose	8.6
Buchan	8.7	Murchison (UK)	10.1
Claymore	10.3	Ninian	9.1
Cormorant Area	10.1	Piper	11.2
Duncan	n.a.	Statfjord (UK)	9.9
Dunlin	10.2	Tartan	7.4
Forties	11.4	Thistle	9.7
Fulmar	10.8		

Source: Kemp and Rose, Aberdeen University

It seems that actual royalty payments in 1984 were made in the following manner:

— BNOC, according to its *Annual Report*, 1984, received an average of 260,000 b/d in royalty in kind in 1984 though it was entitled to receive approximately 285,000 b/d. The difference is due to a number of technical factors. We calculated that the actual liability for fields paying in kind was approximately 232,500 b/d.

— The over-delivery of royalty in kind is compensated for by cash repayments made by the Department of Energy. The sum allocated for this purpose in 1984 was £250 million.[2] Since the

[2] HM Treasury, *Supply Estimates*, 1985–1986

value of the amounts over-delivered (27,500 b/d × 365 = 10 mb) was approximately $300 million, which is close to the £250 million allocation at the exchange rates of 1984, the estimates of effective royalty rates and of over-delivery to BNOC seem to be consistent.

— Payments of royalties in cash for the eight oilfields mentioned in section 8.3 were equivalent to the value of 20,000 b/d. The average effective rates for these oilfields was 8.4 per cent in 1984.

In short, BNOC received more royalty oil in kind than the relevant oilfields were liable for, but the licensees received payments on account throughout the year of the sums due to them for over-deliveries. The comparison of marginal, effective and actual delivery rates for royalty in kind is summarized as follows (percentages of the value of gross production):

Item	Rate (%)
Royalty rate	12.5
Effective rates:	
– oilfields paying in kind	10.2
– oilfields paying in cash	8.4
– all oilfields	10.0
Rate of actual delivery in kind	11.4
Rate of repayment for oilfields paying in kind	1.2

Table 8.2 shows the amounts of net PRT payable in 1984 by oilfields. This concept is defined as follows:

Net PRT payable = APRT + Normal PRT liability − APRT credit − Non-creditable APRT to be repaid after five years.

Thus, the net PRT payable is not always equal to the PRT liability that would normally have been incurred in the absence of APRT, but is adjusted for APRT payments and credits. It is the net sum actually payable after deductions of APRT credits under the combined APRT/PRT regime. Note that the component of 'non-creditable APRT' is introduced here for the sake of completeness and is dropped from Table 8.2 because its value for all listed oilfields is zero.

Table 8.2:   UK. APRT and PRT Liabilities and Credits by Oilfield. 1984.
£ Millions

Field	APRT	Normal PRT	APRT Credit	Net PRT Payable
Argyll	15.7	–	–	15.7
Auk	14.5	–	–	14.5
Beatrice	53.4	–	–	53.4
Beryl	151.9	300.2	151.9	300.2
S Brae	103.3	224.4	224.2	103.3
Brent	595.4	1926.3	595.4	1926.3
Buchan	30.3	40.2	40.2	30.3
Claymore	91.1	258.6	91.1	258.6
Cormorant Area	171.6	–	–	171.6
Dunlin	72.9	169.9	72.9	169.9
Forties	480.4	1924.2	480.4	1924.2
Fulmar	170.2	551.9	201.7	520.4
Heather	27.8	0.2	0.2	27.8
Hutton	5.9	–	–	5.9
NW Hutton	63.1	–	–	63.1
Magnus	112.8	–	–	112.8
Maureen	60.7	–	–	60.7
Montrose	16.9	–	–	16.9
Murchison (UK)	107.5	378.9	107.5	378.9
Ninian	255.4	662.8	255.4	662.8
Piper	233.8	909.8	233.8	909.8
Statfjord (UK)	75.3	188.3	81.1	182.5
Tartan	38.0	–	–	38.0
Thistle	116.8	331.7	116.8	331.7

Source: Kemp and Rose, Aberdeen University

Table 8.2 indicates that eleven fields – Argyll, Auk, Beatrice, the two Cormorant fields, Hutton, North West Hutton, Magnus, Maureen, Montrose and Tartan – had no normal PRT liabilities in 1984. The net PRT payable for these fields is equal to their APRT payment at the rate of 15 per cent as can also be seen in Table 8.3.

Three fields – South Brae, Buchan and Heather – became liable for normal PRT in 1984 or just before. APRT credit is thus set against this liability, and the net PRT payable is again equal to the

Table 8.3:   UK. Effective PRT Rates Defined as Net PRT Payable in Percentages of Gross Revenues. By Oilfield. 1984

Field	PRT %	Field	PRT %
Argyll	15.0	Heather	15.0
Auk	15.0	Hutton	15.0
Beatrice	15.0	NW Hutton	15.0
Beryl	29.6	Magnus	15.0
S Brae	15.0	Maureen	15.0
Brent	48.5	Montrose	15.0
Buchan	15.0	Murchison (UK)	52.8
Claymore	57.4	Ninian	38.9
Cormorant Area	15.0	Piper	58.3
Duncan	15.0	Statfjord (UK)	37.4
Dunlin	34.9	Tartan	15.0
Forties	60.0	Thistle	66.0
Fulmar	45.8		

Source: Kemp and Rose, Aberdeen University

APRT payment due for 1984. The effective rate for these fields is also 15 per cent (see Table 8.3).

For nine fields – Beryl, Brent, Claymore, Dunlin, Forties, Murchison, Ninian, Piper and Thistle – APRT is fully credited and net PRT payable is equal to the normal PRT liability. Statfjord and Fulmar had a small element of past APRT credit so that net PRT payable was slightly lower than normal PRT.

In Table 8.3, effective rates are calculated on the basis of gross revenues to enable us to compare the fields that only pay APRT with the others. Effective PRT rates calculated on the basis of taxable profits are by definition equal to 75 per cent, except in cases where the safeguard provisions still have an impact. According to the Aberdeen model, fields liable in 1984 for normal PRT at an effective rate lower than 75 per cent were:

Field	Effective PRT rate on taxable profits (%)
Brent	62
Claymore	67
Dunlin	63
Ninian	56
Thistle	61

This gives an idea of the continuing impact of the safeguard provisions even on large oilfields.

The North Sea fiscal regime involves a Corporation Tax component. This tax, however, is levied on company profits made on its UK upstream operations, subject to certain provisions mentioned in section 8.5, and not on individual oilfields. It is therefore impossible to estimate Corporation Tax rates by field in any straightforward way using the model. The only calculation that can be made refers to the concept of 'the maximum contribution of the oilfield to the Corporation Tax liability of the licensees'. In other words this concept does not take into account any offsets against this liability arising from the company's operations in other UK fields, and it gives an exaggerated idea of the total tax liability.

The actual fiscal burden is better measured as a share of the net resources generated by the oilfield, in other words in term of its relationship to the field's cash flow. Such measures would show that the fiscal burden is much heavier than is apparent from the data shown here.

8.7   Tax Spinning as an Implication of the Fiscal Regime

The main implications of the UK tax regime for the market behaviour of oil companies relate to a practice referred to as tax spinning. This can be defined as follows. Spinning takes place when an integrated company producing UKCS oil sells part or the whole of its crude oil availabilities in arm's length deals (i.e. in transactions with non-connected companies) rather than appropriating this oil for refining in its own plants, or transferring it to subsidiaries and affiliates. Own crude requirements are then purchased directly from the market for North Sea crudes or from other markets and sources. Spinning may involve different motives, but tax spinning, as the term indicates, aims at reducing the fiscal liability of the integrated oil company operating in the UKCS.

Tax spinning reduces the fiscal liability of an integrated company when the price of an arm's length oil transaction is lower than the price at which a direct appropriation of the oil by the company (or a non-arm's length transaction) is deemed by the tax authorities to have taken place.

It is useful to recall at this point an important feature of the UK oil fiscal system. The UK regime adopts the taxation principle of market or realized prices. In the case of an arm's length transaction (defined in law as a deal between two unconnected parties, in which price is the sole consideration, and which does not involve provisions for resale to a third connected party), the market price is that at which the deal has actually taken place. In the case of a non-arm's length deal (say, between two sister companies of the same corporate group), a direct appropriation or a change in inventory, the equivalent market price has to be ascertained by the tax authorities since there is no visible realization.

The Oil Taxation Office (OTO) therefore seeks to establish this equivalent market price for non-arm's length transactions by referring to the prices of contemporary and comparable arm's length deals. This procedure is perfectly consistent with the 'market price principle' enshrined in the UK oil taxation legislation.

Despite its apparent simplicity this procedure involves difficulties and drawbacks. First, the equivalent market price for non-arm's length deals cannot be established immediately. The OTO must assemble the information about contemporary and comparable deals and then make a decision based on an analysis of the relevant data. The process can be lengthy, and the oil company will be left uncertain for a long period of time about the basis on which taxes will be assessed.

Secondly, the coverage of the concept 'contemporary and comparable deals' is not very clear. The concept may be taken to refer to all market transactions effected at a given time; alternatively to the arm's length deals in which the company itself was engaged at a given time. Furthermore, the term 'contemporary' is itself vague as it could equally refer to a period as short as a day or as long as a fiscal year.

Thirdly, an integrated company which has good reasons to believe that the price at which the OTO will deem an internal transaction to have taken place is higher than the price it can realize in an arm's length deal will be induced to sell its oil output (and buy the oil it requires from other sources) instead of taking the short cut of an internal transfer.

Although the OTO has maintained a consistent view on the principles guiding the taxation of non-arm's length transactions, its views on what constitutes the equivalent market price for these deals have varied over time because of fundamental changes in the circumstances of the oil industry.

Until late 1983 or early 1984, the OTO took the view that a very large proportion of transactions in UKCS crudes were taking place at the BNOC term price, which was therefore taken as the relevant price for taxing non-arm's length deals. This view was justified on the grounds that the spot market was small and could be ignored, and that contract sales, whether involving BNOC or not, were made at prices close to BNOC's.

Sometime towards the end of 1983, or in early 1984, the OTO recognised that the volume of crude oil sold on the spot market had considerably increased. Transactions at BNOC prices could no longer be considered (a) to represent virtually all 'contemporary' deals, nor (b) to be entirely 'comparable'. The equivalent price for non-arm's length transactions could not be equated to the BNOC price without infringing the 'market price' taxation principle. It is legitimate to infer that the OTO then took the view that the valuation of internal transactions of an integrated company should be made on the basis of some weighted average of BNOC, spot and other arm's length prices.

The abolition of BNOC also removes the BNOC term price from the scene. The equivalent market price of internal transactions will have to be ascertained on the basis of spot and contract prices only. The issue, mentioned earlier, of whether the weighted average of these prices should involve all contemporary market transactions or only the transactions of the relevant tax payer, remains to be clarified.

8.8   The Benefits from Tax Spinning

Tax spinning reduces the overall tax liability of an integrated oil company when the following condition is satisfied: the price of the arm's length transaction must be lower than the price at which the OTO assesses non-arm's length deals. This is simply because the overall taxation rate for the upstream operations of oil companies in the UKCS is higher than the taxation rate downstream. (The former includes both PRT and Corporation Tax, and the latter only includes Corporation Tax.)

An overall gain arises only when the difference between the arm's length price and the tax-assessment price exceeds a number equal to the transaction costs involved in spinning the oil (one sale and one purchase) divided by the PRT rate.

Tax gains arise from spinning irrespective of whether the integrated oil company incurs Corporation Tax liabilities in its downstream operations. A tax gain from spinning also arises even if spinning creates a downstream Corporation Tax liability in a situation where direct appropriation does not involve such a liability.

Tax gains arise whether the Corporation Rate of tax is higher or lower than the PRT because in either case the tax rate will remain higher upstream than downstream.

Formal proofs of these propositions are given in Annex 2. The results show that spinning reduces the tax liability of integrated oil companies under all sets of conceivable circumstances so long as the arm's length price is lower than the tax-assessment price (be it the BNOC term price or a weighted average of prices of comparable and contemporary transactions).

However the tax gain has to be set against the additional transaction costs involved in spinning. Simple calculations show that, if transaction costs are 15 cents per barrel, an overall gain from spinning a barrel of oil emerges when the tax-assessment price is 20 cents or more above the arm's length price. Thus, when the OTO used to take the BNOC term price as the basis of its assessments, a spot price just 20 cents lower would begin to provide an incentive for tax spinning.

The simple economics of tax spinning is such as to make the practice almost irresistible. Furthermore, spinning is a perfectly legal activity. The law, after all, does not prohibit arm's length deals by integrated companies! It clearly stipulates that the relevant price for taxation is the market or realized price. It is therefore perfectly legitimate to realize a price for every transaction, if one so wishes. When tax spinning reduces the tax bill of firms, neither evasion nor avoidance is involved. It can be simply said that the firm which spins its oil is organizing its tax affairs sensibly by choosing the type of transactions that attracts both the clearest stipulation of tax laws and the most favourable assessment from the tax authorities.

The significance of spinning does not relate to its fiscal implications (however favourable they may be for the companies concerned) but to its impact on the size, the growth and the behaviour of spot markets. As we shall see in the next section, spinning has supplied the spot market with large and growing volumes of crude oil. This expansion of the liquid base of the spot market increases the very large number of paper transactions through the operation of a multiplier. Sudden increases in the volume of spinning, as in mid-1984 (see section 8.9), are also likely to have some impact on the behaviour of spot prices.

8.9 The Growth of Spinning and the Behaviour of Individual Companies

Integrated companies producing UKCS oil had an incentive to spin their production during seven distinct pricing episodes between the beginning of 1980 and the end of 1984. These episodes are identified in Figure 8.1 which plots the differential between the Brent spot price and the BNOC term price. In general, Brent blend is presumed to be a reasonable proxy for other UKCS crudes and will be taken as representative, though the duration of the periods may have been slightly different for other crudes.

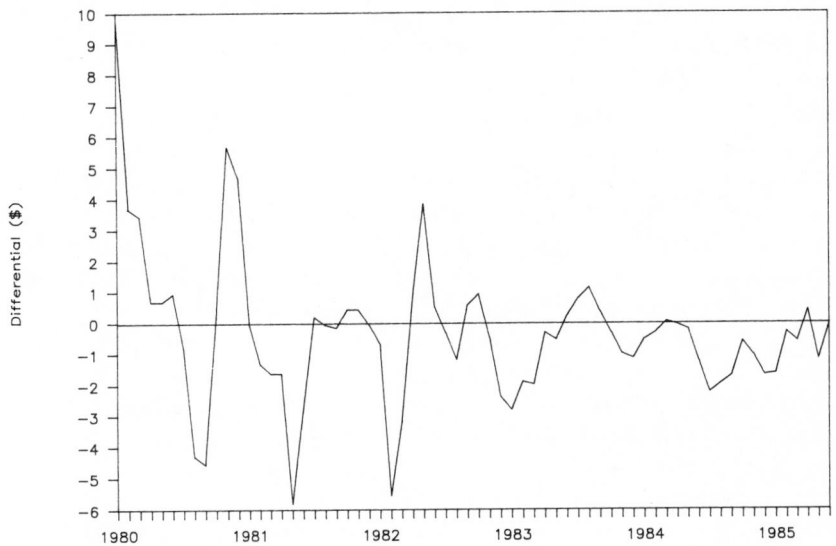

Figure 8.1   Crude Price Differential. Brent Spot/Term January 1980–June 1985

The episodes were as follows:

— July to September 1980
— January to June 1981
— January to March 1982
— July to September 1982
— November 1982 to May 1983
— September 1983 to March 1984
— May 1984 to January 1985

In each case the episode began with a sharp fall in spot price, creating an incentive to spin; and in four cases the episode ended with a price cut by BNOC.

The first Press report on tax spinning was in April 1981[3] and the issue has been regularly discussed since, suggesting that it was regularly practised when the conditions were right. It is said that the fiscal advantage of tax spinning was first perceived by Conoco and almost simultaneously by BP. We do not know whether this is true, but we note that the first reported sales of North Sea crudes by integrated companies at prices below BNOC's included:

— BP to the US SPR, agreed in March 1981, of 55,000 b/d for May and June at $1.25 below the BNOC price[4]
— Conoco to the US SPR of 1 million barrels in June 1981 at $7.25 below the BNOC price

In early 1982, spot prices were again below BNOC term prices; and BNOC was unable to retain all its term customers. The Corporation was left with a surplus of 175,000 b/d which it placed on different terms to the rest of its contracts. The integrated companies began to raise questions about the validity of the BNOC term price as a tax-assessment price for non-arm's length transactions, since it could not be taken as representative of all 'contemporary and comparable' transactions. We can only assume that this provided a spur for spinning in subsequent periods.

In March 1983 BNOC switched from a Forties to a Brent marker price, and put Brent 25 cents per barrel above Forties. Esso was unhappy with this differential, and a public controversy developed which shed light on Esso's attitude towards spinning. According to *PIW* some of Esso's competitors suggested that Esso 'should simply trade its high-priced Brent for lower priced Forties'; to which Esso responded that 'this would probably raise a variety of political and operational difficulties for the company.'[5] In fact Esso's competitors were advising it to spin, and Esso was expressing its reluctance to do so. However, other integrated companies were spinning their oil in different degrees whenever the circumstances were favourable; and there is no doubt that by 1983 tax spinning had become a significant phenomenon.

In mid- or late 1983, oil companies that had previously limited the scope of tax spinning to their equity oil began to seek ways of extending it to the volumes of participation crude that they retained under buy-back agreements. The OTO clarified the point by

[3] *PIW*, April 20, 1981
[4] *PIW*, March 23, 1981
[5] *PIW*, August 15, 1983

restating that a barrel of participation crude bought back by the producing company does not constitute a transaction because no physical delivery has taken place. In case of no delivery, the transaction is ignored for PRT purposes and PRT is chargeable only for later disposal or appropriation of the crude by the company concerned.[6] The door was then open for spinning vast quantities of UKCS oil.

We have attempted to quantify the extent of tax spinning in 1984 and 1985, using proxy measures, in order to describe changes in the patterns of behaviour of the major integrated oil companies. Our estimates concentrate on the four largest export grades – Brent, Ninian, Flotta and Forties – and make use of the tanker tracking exercise mentioned in Chapter 3. The estimates are for third-party sales and therefore overstate the extent of 'pure' tax spinning for companies that are long on crude. Since none of the relevant major oil companies is significantly crude long, the upward bias of the third-party sales measure is likely to be small.

Table 8.4 shows that third-party sales by integrated companies increased markedly as a proportion of their availabilities between the first and the second halves of 1984. The percentage of third-party deals rose from just over 50 per cent to approximately 80 per cent of availabilities. Such an increase in third-party sales by integrated companies at a time when spot market prices had fallen sharply and then remained below the BNOC term price has no reasonable explanation other than tax spinning.

Table 8.4:   UKCS. Availabilities and Third-party Sales by Large Integrated Companies of Major Export Grades. 1984–5. Thousand Barrels per Day

	1Q84	*2Q84*	*3Q84*	*4Q84*	*1Q85*
Availabilities	1192	1126	1118	1227	1244
Third-party sales	674	546	877	1006	792
Ratio (%)	57	48	78	82	64

However, individual companies did not always behave in the same manner. Table 8.5 compares third-party sales of BP, Esso and Shell, and shows that:

[6] House of Commons, *Fourth Report from the Energy Committee*, Session 1984–85, p34.

— BP and Shell tended to respond to market circumstances. They spun larger proportions of their oil in 1Q84 when the spot price was below BNOC's than in 2Q84 when the spot price was at times above BNOC's; and they increased these proportions substantially in the second half of 1984 when the gap between spot and BNOC prices was very large.

Table 8.5:   Availabilities and Third-party Sales of Major Export Grades by BP, Esso and Shell. 1984–5. Thousand Barrels per Day

	1Q84	2Q84	3Q84	4Q84	1Q85
BP					
Third-party Sales:					
Ninian	73	115	135	130	138
Forties	230	142	173	231	227
Total	303	257	308	361	365
Availabilities:	401	457	421	445	448
Ratio (%)	76	56	73	81	81
Esso					
Third-party Sales:					
Brent	53	80	170	178	57
Availabilities:	305	249	253	294	305
Ratio (%)	17	32	67	61	19
Shell					
Third-party Sales:					
Brent	169	69	176	237	171
Availabilities:	305	249	253	294	305
Ratio (%)	55	28	70	81	56

— Esso seems to have been influenced more by strategic objectives than by purely market factors. The proportion of third-party sales was very small in 1Q84 but increased in 2Q84 when there was no fiscal advantage to be gained from spinning. Like BP and Shell, however, they significantly increased their tax-spinning activity in the second half of 1984; and unlike BP and Shell they virtually ceased to spin in early 1985 though the fiscal advantages continued to prevail.

— Finally, it appears that on average BP disposes of a larger proportion of its oil availabilities in third-party sales than Shell. In the second half of 1984, however, the ratio of third-party sales to availabilities was virtually identical for these two companies. Esso sells a much lower proportion of its oil to third parties than either of its two Sisters. There is clearly a fundamental difference in approach to oil trading and supply between Esso, on the one hand, and BP and Shell on the other.

8.10 Conclusions and Implications

From the point of view of an integrated oil company, spinning is tax efficient whenever spot prices happen to be below the price at which the OTO is likely to assess their fiscal liabilities (be it the BNOC price, a weighted average of the BNOC and spot prices or any weighted average of prices).

Some commentators mistakenly believe that, since the tax gains from spinning are higher the greater the gap between spot prices and the tax-assessment price, oil companies have an interest in pushing the spot price down. Things are not so simple. For such a strategy to succeed, oil companies must wield power in both crude and oil product markets. In order to maximize gains from tax spinning, the spot crude price should be depressed and the oil product price sustained. The majors may have some influence on the crude oil market but it is certain that they have no monopoly power in the product market.

Oil companies, though perfectly aware of this point, may however underestimate the extent of their *collective* power. A sudden increase in oil supplies on a particular spot market, due to an increase in tax spinning, is likely to affect the behaviour of spot prices because these additional supplies may not be matched immediately by a commensurate expansion of demand on that market. Whenever spinning brings down the spot price, tax gains could be more illusory than real. They could be easily cancelled, or even outweighed, by losses arising from the fall in product prices.

Since such a fall induces more tax spinning, a silly chase for gains which the chase itself wastes away can take place. One may question the wisdom of it all while conceding that an individual company has no choice but to conform to the common behaviour, because the actions of others continually make a new spot price which becomes a given datum for that company. It is easy to understand both Esso's reluctance to spin its oil and its recourse to spinning when the actions of others deprived it of any other option.

The abolition of BNOC raises the question of whether tax spinning will continue or abate. The answer partly depends on companies' perceptions of the relationship between spot prices and the prices at which the OTO will value their non-arm's length oil transactions. If companies remain uncertain about how they are to be assessed, they may continue to spin – not necessarily to realize a tax gain but to reduce the uncertainty. Spinning will become a sort of fiscal hedging turning a future uncertainty (the *unknown* price the OTO will use for its assessments) into a present certainty (the *known* price of the arm's length transaction undertaken today).

If the OTO assessment of non-arm's length transactions were made on the basis of an average of spot prices over a relatively long period of time (say, a month or a quarter), companies might be tempted to speculate against the expected average price. They would spin when they felt that spot prices were below their forecast of the monthly (quarterly) average and refrain from spinning when they thought that spot prices were above the expected average. This is not a far-fetched proposition. Traders continually take views about spot price movements and could easily apply their skills to this new form of spinning.

Finally we believe that, even if the motivation for tax spinning weakens, some integrated oil companies will continue to dispose of a proportion of their oil in third-party sales. Of course, this proportion will be lower than it is at present but could remain significant. The reasons are threefold:

— First, trading departments have become powerful and dynamic entities within oil companies. They are strongly motivated to maintain a high level of trading activity.
— Secondly, many corporations have granted considerable commercial autonomy to their subsidiaries. These changes in corporate structure are unlikely to be reversed. Subsidiaries may not always wish to buy oil from within the corporate group. The producing subsidiary might have no option but to sell its oil to third parties, and the refiners within the group would then continue to purchase oil from a variety of other sources.
— Thirdly, much third-party trading is justified by integrated oil companies on the grounds that they need to shop around continually to supply their refineries with the most suitable crudes obtained at the best possible prices.

To sum up; tax spinning has probably been instrumental in the development and growth of spot markets for North Sea crudes.

These markets now have a life of their own. A weakening of the motivation for tax spinning would undoubtedly reduce oil supplies to the North Sea spot markets, but would probably not reduce them to an insignificant level or destroy their momentum.

CHAPTER 9

NORWEGIAN LICENSING AND PRODUCTION POLICIES

9.1 The Norwegian Regulatory System and its Political Framework

The Norwegian Continental Shelf is being explored and developed by a partnership of private enterprise and the state-owned company, Statoil. Since 1973 Statoil has been awarded a minimum of 50 per cent interest in each production licence. The fact that, especially in the earlier rounds, it has not been the operator despite being the largest shareholder in a block has made this partnership at times an uneasy one.

The institutional framework for the petroleum industry was defined by the Royal Decrees of 1963, 1965 and 1972 which were introduced by Labour and left-wing Coalition Governments. Statoil was established in 1972 as a vehicle for state participation in the industry and given very general guidelines which left it free to find its own role.

The Conservative Coalition Government, which came to power in 1983, has made no change to the legislation covering offshore activities. Indeed, the philosophy that has governed the development of Norway's oil resources from the early 1970s to the present day seems to be broadly shared by politicians from all parties, who share the belief that the impact of the petroleum industry on the Norwegian economy is too great to allow private enterprise to have complete control over the depletion of offshore resources.

The role of Statoil, however, has come under scrutiny from the Conservative Government. There were fears that the substantial cash flows that will accrue to Statoil in the late 1980s, when the Gullfaks field comes into production, would make it too large and powerful a force in an economy as small as Norway's. Statoil's traditional role as equity holder in producing fields is therefore to be modified so that it remains the seller of the petroleum but not the recipient of most of the cash flows. The new arrangements will come into force in 1987 when Gullfaks is due to come on stream. Statoil's equity interest in the field will be reduced from 85 to 15 per cent with the remaining 70 per cent being held directly by the state. The crude oil accruing to the state will be marketed on its behalf by Statoil.

The fiscal regime introduced by the Socialist Government has been left largely unchanged by the Conservatives with the exception of the concessions made to allow the development of the Ekofisk Waterflood Project (see section 10.8). They also set up a committee to examine alternative tax structures that might be used in the future to increase the incentives to develop new fields. This committee had no remit to consider a reduction in the level of taxation, which has remained consistently high under all Governments.

9.2   The Licensing of Exploration and Production

Sovereignty over the Norwegian Continental Shelf regarding exploration for and exploitation of natural resources was vested in the king by the Royal Decree issued in May 1963. This gave the king powers to issue regulations governing such activities. The Royal Decree of April 1965 (later superseded by the Royal Decree of December 1972) established the main guidelines for Continental Shelf activities, including the right of the Crown to grant licences for the exploration and production of petroleum.[1] Reconnaissance licences give permission to carry out seismic surveys and other types of research, but not to drill. Production licences give the licensees exclusive rights to explore for and produce petroleum in the licensed area without any limit to the depth of drilling.

The choice of areas to be opened and blocks to be awarded is based on evaluations by the Norwegian Petroleum Directorate. The awarding of production licences takes place after bilateral negotiations between the Ministry of Petroleum and Energy and the applicants. There is no auctioning involved. The discretionary system of awards is preferred by the Norwegians since it allows them to have a high degree of state participation and to ensure the positions of the Norwegian companies, Norsk Hydro and Saga. It also gives them flexibility in the selection of companies, and enables them to specify conditions, such as requiring comprehensive geological information from licence applicants.

There have been nine rounds of Norwegian production licences

---

[1] The Royal Decrees of 1963 and 1972 are reprinted in *Legislation concerning the Norwegian Continental Shelf with unofficial English translation* (5th edn. 1977) published by the Royal Ministry of Industry and Handicraft, Oslo.

to date. A tenth round was announced in February 1985 inviting applications for forty blocks for which licences are being awarded in late 1985 and early 1986.

Six of the nine rounds have taken place since 1979, the Government having implemented an unofficial 'go slow' policy in earlier rounds in order to prevent production reaching too high a level too soon.

The first round, held in 1965, was the most extensive there has been in Norway, offering 278 blocks and awarding twenty-two licences to cover seventy-eight blocks. Each licence covered a maximum of four blocks and was conditional upon the licensees undertaking a minor work programme and agreeing to pay royalty on any commercial discovery made. No state participation was required but some preference was given to consortia that included Norwegian companies. The largest discoveries on blocks awarded in this round were the fields in the Ekofisk area and Valhall.

The second round, which took place in 1969–71, was more restricted than the first, offering sixty-eight blocks and awarding only fourteen. Each licence covered only one block. A limited form of government participation was introduced in this round: net profit sharing agreements were reached with licensees on six blocks (including the North East Frigg and Odin gas fields) and the state took carried interest options in other blocks, two of which (Frigg and Heimdal) have proved commercial.

Subsequent rounds differed from the first two in that Statoil had by then been established as a vehicle for state participation on the Continental Shelf. The award of licences in the third round, which took place between 1974 and 1978, was conditional upon Statoil taking a 50 per cent interest in all concessions, with an option to increase this up to 70 per cent in the event of a commercial discovery. Operators' tasks were also assigned to Statoil, Norsk Hydro and Saga. Technical assistance agreements with the large oil companies were entered into in these cases to enable the Norwegian companies to gain experience as operators. Royalty payments for oil in blocks awarded in this round changed from a straight 10 per cent to a sliding scale ranging from 8 to 16 per cent. Thirty-two blocks were offered in this round; eight were awarded in 1974 and a further twelve in the period 1976–8.

The conditions attached to the award of licences in the fourth round in 1979 were more stringent and form the basis of all subsequent licence agreements. These covered Statoil participation and more state control over the development and production·

profiles of commercial fields. Out of the fifteen blocks offered in this round only eight were awarded.

The fifth and seventh rounds in 1980–82 allocated seventeen blocks north of 62°N, which marks the edge of the Norwegian Continental Shelf in the North Sea. Three main areas, Halten-banken, Trænabanken and Troms, are currently being explored. The sixth round in 1981 comprised nine blocks, or portions of blocks, in the southern North Sea covering acreage awarded in previous rounds which had subsequently been relinquished. The eighth and ninth rounds in 1984–5 allocated a total of thirty blocks, eleven of which were in the North Sea and the remainder in the Northern blocks.

In addition to the licensing rounds, four special awards of blocks with high commercial potential have been made by the Ministry. The first of these was the award in 1973 of the two blocks adjacent to the recently discovered UK Brent field to a Statoil/Mobil consortium. The Statfjord field lies in these blocks. The other three special awards in 1978, 1982 and 1983 all went to a consortium of Statoil, Norsk Hydro and Saga. These awards comprised the Gullfaks field and blocks adjacent to the Oseberg and Troll fields.

The tenth round is being awarded in two phases: the first covered eight North Sea blocks which were awarded in September 1985 and the second covers thirty Northern blocks to be awarded in the first half of 1986.

9.3   Government Control Over Oil Production

The Norwegian Government has instituted very strict powers over the production level of petroleum on the Norwegian Continental Shelf. Depletion policy has been widely debated in Norway since the development of Ekofisk in the early 1970s and detailed legislation has been introduced enabling the Government to implement this policy through its control over each stage of the exploration and production activity. However, a clear distinction should be made between policy objectives and policy implementation. Despite the state's very wide powers, the Norwegian approach, like that in the UK, has been to exercise indirect control over production levels through the limiting of exploration activity and the conditions attached to the award of production licences.

Depletion policy was first extensively discussed in Norway in 1974 in the Report No. 25 to the Storting, *Petroleum Industry in Norwegian Society*. This was a far-reaching report on the effects of the expanding oil industry on many aspects of Norwegian society. It covered the direction of operations in the North Sea, control of the production rate, possible effects on industry and banking, environmental consequences to the fishing industry, the role of oil and gas in Norwegian energy policy, public revenue from the oil industry and the effect on the labour market.

The Report concluded that Norway should maintain a 'moderate rate of extraction' of petroleum resources, this moderate rate being defined as an annual production level of 90 mtoe (including gas) throughout the 1980s.[2] Much of the discussion centred on the means by which this extraction control should be implemented. The main recommendation was that exploration activities should be regulated according to the size of the discoveries already made since 'once discoveries are made, technical, economic and political reasons make it difficult to limit their exploitation.'[3] The Norwegian Government effectively followed this advice since only three licensing rounds were held between 1965 and 1978, in which only 115 blocks were awarded. In contrast, the UK held five rounds in this period with the award of 907 blocks.

However, the Report recognised the uncertainties inherent in relying entirely on the control of the exploration activity, in particular the difficulty in maintaining an even level of development and production activity with the consequent impact on Norwegian industry and employment. Three possible methods of control after a commercial discovery had been made were suggested: regulating the production profile, delaying the start of production, and increasing the role of the state (through Statoil) in the production process. These suggestions were fully incorporated into the conditions for award of fourth round production licences in 1978–9 which stipulated that:

— Statoil was entitled to an initial carried interest of 50 per cent in each block; this interest could increase to a stipulated maximum (70–75 per cent) according to the estimated maximum production from the field. It was to pay no exploration costs.
— In the event of a commercial discovery, a production profile had to be agreed with the Ministry.
— In some cases the development of a field could be delayed for up to five years.

---

[2] Royal Norwegian Ministry of Finance, Parliamentary Report No. 25 (1973–74), *Petroleum Industry in Norwegian Society*, pp6*, 16* and Appendix p17.
[3] *ibid*, p6*.

The intention of these provisions was to ensure that the authorities could regulate the level of total petroleum production and co-ordinate current development projects to achieve an even level of activity and to provide as much stimulus to Norwegian industry as possible.

The above conditions were extended in the fifth round to allow commercial development to be delayed indefinitely.

By the 1980s the Government therefore had full control over each stage of activity on the Continental Shelf. It could regulate the rate of exploration through the licensing rounds, regulate the timing of development through the licence conditions, and influence the production profile.

By virtue of its voting power and, in some cases, its status as operator, Statoil was in a position to control field development and production profiles to the state's best advantage. In addition, the Royal Decree of 1972 gave the state the power to halt all activity at any stage in exploration or production if it considered that the country's interests would be best served by doing so.

In fact the target figure of 90 mtoe of oil and gas per year has never been reached. Production in 1984 was around 55 mtoe and this is expected to reach 60 mtoe by the end of the decade. In the early 1980s discussion in Norway centred more on how to develop a policy to give 'maximum flexibility' to the Government. In 1983 a Royal Commission reported on the consequences to Norway of alternative levels of production and activity on the Norwegian Continental Shelf.[4]

The Commission suggested that rather than using a tonnage figure as the basic criterion, the Government should look at the relationship between, on the one hand, the income to the state from petroleum activities and, on the other hand, the country's total consumption and investment. In this way the petroleum activities would be linked to criteria that would show more clearly the consequences to Norway's economy of the level of petroleum production. The Commission emphasised as particularly important the need to shield the internal economy from the short-term swings of petroleum revenues. One means to achieve this would be to accumulate a fund which could serve as a buffer for such swings of activity and income.

Since 1978 six more licensing rounds have taken place, and the

---

[4] See summary, in translation, of the White Paper of 19th October 1984, based on the recommendations of the Commission; available from Royal Ministry of Petroleum and Energy.

tenth was announced in February 1985. This follows a recommendation made in a Report to the Storting in 1980 that licensing rounds should be small but should take place on a more regular basis. An increase in exploration activity, coupled with the powers to delay development, may give the Government greater flexibility in choosing which fields to develop and when. If successful, it will also give them the option to increase the overall level of production in the future.

In conclusion, the Norwegian Government has wide-ranging powers to control the rate of production on the Continental Shelf. Nevertheless, the restrictive awarding of production licences and increasing state participation through Statoil are the only means by which it has exercised this control. This is not to say that their policy has been to allow private industry to dictate the rate of depletion, but rather that the interests of the companies and the state have so far coincided. With the decline in Ekofisk production, the state will in fact control more than 50 per cent of Norwegian production by the end of this decade, both through direct participation and through the activities of Statoil.

9.4 The Role of Statoil in the Institutional Framework

The decision to form a state oil company was adopted unanimously by the Norwegian parliament (the Storting) on 14th June 1972. The company was named as Den norske stats oljeselskap A/S and is known generally as Statoil.

Following the first commercial discoveries on the Norwegian Continental Shelf in 1969, various means of state participation in the petroleum industry had been discussed in Norway. The second licensing round introduced options for state participation either as net profit interest or in the form of a carried interest share. The reasons for establishing government participation were two: first, to increase income from petroleum production and secondly, to allow the Government to influence decisions made by the individual licensees. It was argued that, although revenues could be increased without direct Government participation, the Government was interested not only in income but also in the manner in which the country's natural resources were utilised. The impact of a growing petroleum industry on a country the size of Norway was likely to be substantial and the Government wished to exercise as much control over its development as possible. Thus Statoil was formed as the commercial arm of the state in the oil industry.

The legislation covering Statoil is imprecise and its objectives are general rather than particular. This has allowed Statoil the freedom to determine its own role in the industry. Various duties outlined for Statoil in the original legislation include taking an active part in petroleum exploration and production, developing refining and marketing activities, and co-ordinating the Government's interests with the interests of private enterprise through co-operation agreements. Statoil was conceived as an integrated company with interests in each stage of petroleum exploitation.

The main role of Statoil has always been in the upstream side of the business. On its formation the state's participation agreements in the Frigg and Heimdal fields were assigned to it, and all subsequent licensing rounds carried conditions that gave Statoil an initial 50 per cent equity interest with an option to increase its participation on a scale based on the eventual peak production from the field. Its exploration costs were carried by the other licensees. Since 1983, Statoil has covered part of its own exploration costs when in partnership with other Norwegian companies. When a field is declared commercial Statoil covers a share of development costs corresponding to its ownership share. Since Statoil finances the most substantial part of development costs, it has, in most cases, a majority voting right in each field, which allows it to control each stage of exploration and development.

The Government's aim of 'Norwegianizing the petroleum industry' has been furthered by assigning operator tasks to Norwegian companies, including Statoil. Technical assistance agreements with foreign oil companies were entered into in these cases to enable the Norwegians to gain the necessary experience and competence to manage exploration and production activities on their own. Statoil has been appointed operator for the Gullfaks field, due on stream in 1987, and is expected to take over operatorship of Statfjord from Mobil in the same year. It also owns majority shares in the two pipeline systems, Norpipe and Statpipe.

Statoil's downstream activities centre around its 70 per cent share in the Mongstad refinery which came into operation in 1975 and is currently being expanded to a capacity of 6.5 mt per annum. The products are marketed in Norway by its subsidiary, Norol. Statoil also owns one-third shares in two petrochemical companies, Noretyl and Norpolefin.

In 1984 Statoil had a primary availability of crude oil of 270,000 b/d (13.1 mt), of which 94,000 b/d (4.5 mt) was royalty oil bought from the state at norm price (see section 9.6). Statoil's equity production from the Statfjord field covered 60 per cent of its availability. The importance of Statoil's role as a seller of crude can be seen from the fact that it currently markets 40 per cent of Norway's production and by the end of the decade will account for 60 per cent. It will then be the biggest seller of crude in the North Sea. Very general guidelines have been laid down for the marketing of crude oil by Statoil. These are:

— that its task is 'to realize the value of the State-owned crude oil supply.'
— that emphasis should be given to ensuring profitability over a long period by giving priority to safety of disposal over short-term profit in the spot market.
— that sales should be apportioned to ensure stable off-take, spread risks and maximize short-term flexibility.[5]

Statoil's marketing policy has followed these guidelines closely. At the same time it has used its commercial independence to negotiate selling prices in order to retain its term customers in a difficult market. Indeed it has been notably successful in this, selling 80 per cent of its crude under term contract at a time when refiners are increasingly turning to the spot market. Term customers are important to Statoil, first because of the substantial amount of offshore-loading crude available to it which cannot be easily disposed of in the spot market, and secondly because its crude availability will increase substantially over the next decade. It was for these reasons that the company began discounting its official selling prices in October 1984, a move which was blamed for the general slide in prices in the following months. In fact Statoil claims to be a price follower rather than a price leader. The official selling prices, which it posted up until September 1984, were closely based on BNOC official prices, with adjustments for quality and transportation. Rather than follow BNOC into being largely a spot or short-term contract seller, Statoil instigated a policy of negotiating individually with each customer. Its contracts are volume-based and prices are negotiated periodically. Statoil perceives its advantages to European refiners as a secure near-at-hand source of good quality crude, well suited to the product slate of most countries on the North Sea rim. Most of its sales are into this region. In 1984 only 12.5 per cent went to North American markets, with 37 per cent going to Scandinavia and the remainder to North West Europe.

[5] Ministry of Petroleum and Energy, Storting Report No. 53 (1979–80), *Concerning the Activity on the Norwegian Continental Shelf*, pp87–8.

Constitutionally, Statoil is answerable only to the Minister for Petroleum and Energy and has no direct political responsibility or formal connection with the Storting in the conduct of its business affairs. The amount of political control that should be exercised was discussed in a Report to the Storting shortly after Statoil's formation. It recommended that the company should have full powers to take commercial decisions, 'however large, intricate or far-reaching they may be.'[6] The justifications for this were that:

— For Statoil to function effectively as a commercial entity it must be seen to be free from political constraints.
— If every decision had to be referred back to the Ministry, the decision-making process would tend to become very slow and cautious, at the expense of Statoil's business activities.

Large investments had to be approved by the Storting, but since these investments were in offshore development the debate was not whether they should be funded but rather by what means. Statoil's large capital expenditure programme was covered by share capital and direct loans from the state, along with external loans carrying a state guarantee.

For the first ten years of its existence Statoil expanded its activities unhindered by its political masters. The election of a Conservative Coalition Government at the end of 1981, however, signalled the end of this period of uninterrupted growth. The new Government's election coincided with the first substantial cash flows from the Statfjord field. With this large income and its majority voting rights in developing fields, Statoil was becoming a powerful influence in the Norwegian economy. The growth in its cash flow enabled it to act increasingly independently – more like a private company than a state-owned organization in the view of some politicians. The Government determined to curb Statoil's growth within Norway. Instead of following the path chosen by the UK Government of privatising the upstream business to direct the cash flows into shareholders' pockets, the Norwegian Government has chosen to take more of the cash flow directly into the Finance Ministry instead of allowing it to accrue to Statoil. Licences awarded in the ninth round still include a 50 per cent share for Statoil in each block, but its voting rights have been reduced to 20 per cent, with the state holding the remaining 30 per cent directly. The

[6] The Ministry of Industry, Report No. 30 to the Norwegian Storting (1973–74), *Operations on the Norwegian Continental Shelf etc.*, p195.

Government has also reduced Statoil's equity share in Gullfaks from 85 to 15 per cent and will hold the difference itself.

The results of these changes will be much smaller cash flows and less influence for Statoil. Its role as a crude oil seller will remain unchanged, however, since it will still market the state's share of petroleum. In return for reducing its influence within Norway, the Government is now allowing Statoil to expand its activities abroad. It has recently acquired downstream networks in Sweden which will provide an outlet for the products from the Mongstad refinery expansion. The expansion and the moves into retailing are part of a longer-term plan aimed at giving Statoil more flexibility in disposing of the large amounts of crude that will be available to it by the end of the decade. Its ambition certainly seems to be to become a powerful European integrated refiner rather than a Norwegian producer. This could herald the start of a new period of growth for Statoil.

CHAPTER 10

NORWAY. THE OIL FISCAL REGIME

10.1 Introduction

The Norwegian Government derives its revenue from offshore activities from four sources:

— Royalty payments based on gross field revenues. Royalties are currently paid in kind but may be paid wholly or partly in cash at the Government's discretion.
— Special Tax (ST) based on gross company income derived from exploration, production and pipeline transport on the Norwegian Continental Shelf.
— Corporation Tax based on total company profits. For profits deriving from Continental Shelf activities (as for ST) all Corporation Tax is paid to the state; for other profits, a proportion of this tax is paid to the municipality. The distinction is important since distributed dividends are deductible from national tax but not from Municipal Tax.
— Capital Tax based on the written down value of capital equipment.

These various taxes will be described in the following sections along with a description of the norm price system which is used to calculate gross income for the purposes of Special Tax and Corporation Tax. The norm price is an administered price; the pricing principle of the Norwegians for their oil fiscal regime therefore differs from the UK 'market price' principle.

10.2 Royalties

The legal provision for the payment of royalties was first set out in the Royal Decree of 9th April 1965 and subsequently revised in the Royal Decree of 8th December 1972.[1] These contained detailed

---

[1] Royal Ministry of Industry and Handicraft, *Legislation concerning the Norwegian Continental Shelf with unofficial English translation* (5th edn. 1977).

legislation covering all aspects of offshore exploration and production in Norway. The original terms apply to all licences awarded before 1972, i.e. in the first two rounds of licensing. As in the UK, royalty is deducted before tax liability is assessed.

Royalties are levied on the total production of each field and are based on well-head values. The amount levied depends on the round under which the production licence was issued. The position is as follows:

— Licences issued in rounds 1 and 2: royalty is 10 per cent of total production.
— Licences issued subsequently: royalty is paid on a sliding scale depending on the daily average production of the field over a 30-day period.

Production	% Royalty
less than 40,000 b/d	8
40–100,000 b/d	10
100–225,000 b/d	12
225–350,000 b/d	14
more than 350,000 b/d	16

If production from the field is reduced over a 30-day period, the royalty is reduced accordingly. However, if the production level falls below the 100,000 b/d threshold, the royalty rate remains at 12 per cent.

If a field extends over several licence areas, the royalty rate is determined by the total production of the field on the sliding scale.

The Government has the power to stipulate a lower rate of royalty or to remit payments altogether for a particular field if conditions so require. This power has never been used but could be invoked if, for example, the high royalty rate of 12 per cent payable on a declining field were to lead to that field being prematurely abandoned as a result of operating costs exceeding revenues after payment of royalty.

Royalties are calculated as a percentage of gross revenues less allowable costs. Gross revenues are calculated as norm price × (gross production less petroleum used for production purposes at the place of production). The only deductions allowed against gross revenue are:

— area fees (payable on production licences after the first six years)
— costs of pipelines, loading buoys and tankers
— initial treatment and transportation costs (negotiable with the state)

Royalties can be paid either in cash or in kind at the discretion of the Minister of Petroleum and Energy. Since March 1974, all royalties have been paid in kind. The Ministry may, with six months' written notice, require royalty to be paid wholly or partly in cash. If paid in cash, it is payable quarterly within thirty days of the end of the period and the value of the oil is calculated for royalty purposes on the basis of the norm prices determined for that quarter (see section 10.7 for a description of the norm price system).

When payments are made in kind, a monetary adjustment is required to allow for deductible costs, since the full percentage rate is applied to physical output. The royalty oil is received by Statoil on behalf of the state at the production area shipment point. Statoil purchases the royalty oil from the Government at the norm price and is then free to dispose of it as it wishes. The fields in the Ekofisk area (including Valhall) pay the flat 10 per cent rate while Statfjord and Murchison are assessed on the sliding scale rate.

10.3 Special Tax (ST)

This tax has its origin in the Petroleum Taxation Act (PTA) of 1975 which was introduced following the surge in oil prices of 1973–4 and the consequent increase in profits of companies operating on the Continental Shelf.[2]

Special Tax, unlike PRT, is non-deductible before Corporation Tax, but, like PRT, applies only to income derived from exploration and production activities. However, there are differences both in the method of gross income evaluation and in the definition of allowable expenses. The most important of these is that ST is levied on companies rather than on a field-by-field basis; however a partial ring-fence does exist in Norway owing to the provisions for offsetting the capital cost of field development.

A description of the rules for ST assessment follows.

The assessable profit for ST purposes is the gross income derived from petroleum production and pipeline transport on the Norwegian Continental Shelf less allowable operating and capital expendi-

[2] *ibid*, pp85–97.

ture (including uplift), interest charges and losses carried forward from offshore business.

Gross Income. This is defined as all income derived from petroleum production and pipeline transportation on the Norwegian Continental Shelf. Proceeds from sales of petroleum (after deduction of royalties) are taxed not on an invoiced price basis but by a series of tax-reference prices (norm prices) prescribed by the authorities (see section 10.7 for a description of the norm price system). Profit on the sale of disposable assets is taxable as income earned from offshore activities. Income derived from the transportation and treatment of 'third-party' oil – that is, oil that is not owned by the licensees – is also subject to Special Tax. Some companies that hold production licences have raised loans to finance the development of fields on terms stating that the lender shall receive part of his remuneration in kind, i.e. produced petroleum. A supplement to the 1975 Act states that remuneration such as a share of the production or profit is liable to Special Tax.[3]

Royalty. As in the UK, royalties paid in kind are by definition excluded from the gross income calculation since no proceeds are derived from them.

Expenditure. Deductions for offshore expenditure for each company are allowed subject to certain provisions. Exploration costs and operating costs incurred in the development of other fields can be set against income from producing fields, as can interest payments on loans acquired for the purpose of financing offshore activities. Development costs, however, must be capitalized, and are allowable as deductions only in the form of annual depreciation. Depreciation can be charged on a straight line basis over six years (16.67 per cent per annum). However, allowances may only be taken from the time that the equipment comes into 'ordinary use', i.e. from the start of production. This provision has a similar effect to a ring-fence round each field since the capital costs of a new field may not immediately be set against the income from producing fields. Once production has commenced, however, any deficit incurred in the field in question (including depreciation) may be deducted from any other income the producer may incur from the Norwegian Continental Shelf. Thus no complete ring-fence system has been established.

The 16.67 per cent depreciation rate is a maximum rate and

[3] Royal Decree of 25th June 1976, in *ibid*, pp177–187.

companies may choose a lower one. However, they may not change their depreciation rate at will from year to year.

Deficits arising from offshore activities may be offset against any other income earned on shore. The tax payer may, however, choose to carry forward the deficit against subsequent income subject to Special Tax even if he has a profit on shore. A petroleum production deficit may be carried forward for fifteen years, but it is possible to extend this time-limit. There is no limit on the level of losses that can be carried forward in a single year. Deficits arising from onshore business are not deductible for Special Tax purposes.

Uplift. Permanent installations offshore employed in petroleum exploration and production activity qualify for a special allowance known as uplift. It is deductible only against Special Tax. Other plant and equipment onshore are excluded.

In 1975 the uplift was set at 10 per cent over fifteen years; in 1980 this was reduced to 6.67 per cent, i.e. a total of 100 per cent of investment. Uplift does not come into effect until one year after production from the field has actually begun. If the uplift exceeds regular taxable income, the excess may be carried forward indefinitely and set against future years' income.

Uplift also extends to pipelines that transport petroleum produced on the Norwegian Continental Shelf. Where third parties are involved, uplift is prorated on the basis of use rather than ownership, unless the pipeline is owned by a company whose sole business is to rent pipeline capacity. In this case the uplift is allowed as a cost against income. If the uplift should exceed the company's income, then the users share the excess proportionately.

When equipment is sold the proceeds are used to reduce the uplift up to the limit of the original cost of the item. If equipment is transferred to another, non-qualified, user the uplift basis is reduced by a formula:

$$\frac{\text{The remainder of the 15-year period} \times \text{original cost}}{15}$$

If the production period of a field is shorter than fifteen years, any unutilised uplift may be carried over and used to calculate the base for ST on other fields. In 1980, an amendment was made to the Petroleum Taxation Act under which a tax payer may claim reassessment on termination of production if he has not been allowed full uplift and if there is no other Special Tax income from which he may claim uplift. When reassessment takes place, deductions in Special Tax income are allowed so that the uplift amounts to 100 per cent of the cost of depreciable assets on the field, i.e. the annual uplift deduction is increased.

The Special Tax rate on taxable income is currently 35 per cent. The previous rate (up to the end of 1979) was 25 per cent.

10.4 Corporation Tax

Corporation Tax in Norway is charged at a rate of 50.8 per cent on gross income. For onshore business 23 per cent is a Municipal Tax and 27.8 per cent a national Income Tax. Tax on offshore income is in practice all paid to the state, but the distinction between the two rates remains for the purposes of evaluating allowable deductions on dividends. Corporation Tax is charged on the same gross income as Special Tax.

In addition to the allowances described in the previous section, deficits arising from onshore business may be deducted for Corporation Tax purposes up to a maximum of 50 per cent. A deficit incurred from business outside Norway is not deductible.

Dividends declared payable to the shareholders are deductible against the national tax portion of Corporation Tax. The allowable deduction for dividends in computing the national tax is applied against income for the preceding year out of which the dividend is paid. If a company distributes dividends in excess of taxable profits, the deduction is limited to taxable income for state tax purposes; i.e. the national tax payment may be nil, but the company cannot obtain a credit or refund.

10.5 Withholding Tax and Double Taxation

When a dividend is declared, a special Withholding Tax is paid when the distribution is made by a Norwegian subsidiary to its foreign parent. The standard rate of Withholding Tax is 25 per cent but where a double taxation agreement exists between Norway and the foreign country concerned this rate may be reduced. The precise rate depends upon the degree of ownership by the foreign parent, with the rate of tax being lower the higher the foreign ownership. The rate currently levied on US and UK subsidiaries is 15 per cent.

Only domestic companies (including subsidiaries of foreign companies) are eligible for dividend deductions. The Norwegian branch of a foreign company cannot distribute dividends. However, the double taxation agreement between Norway and the USA contains a non-discrimination clause which essentially allows a Norwegian branch of a US company to claim deduction for its proportionate share of dividend distributed by its parent. The US branch pays tax of 15 per cent on this dividend, thus achieving a net tax saving of 12.8 per cent on the dividend amount. The tax treaty between Norway and the UK states that Norway has no obligation to allow deduction for dividends for British companies.

10.6 Capital Tax

Capital Tax is levied at a rate of 0.6 per cent on the written down value of a company's capital equipment. This is not deductible for either Income or Special Tax.

10.7 The Norm Price System

The 1975 Petroleum Taxation Act stipulated that, for the purposes of determining gross income for Special Tax and Corporation Tax, the value of petroleum extracted from the Norwegian Continental Shelf should be determined administratively rather than from actual invoiced prices. The reason for this decision was the difficulty envisaged by the Government in determining a value for oil that is not traded on the open market. In the White Paper to the 1975 Act it was pointed out that much of the crude oil in international trade is transferred among affiliates of integrated oil companies; the transfer prices are more likely to be influenced by the companies' risk-spreading across the group than by market forces. Transactions between non-associated companies may also be linked to other transactions in other parts of the world (exchange deals). Other factors, such as agreements with state oil companies, production restrictions and price regulations, make the assessment of taxable income on the basis of invoiced prices a particularly difficult and complex system. The norm price system was introduced as a means of simplifying the tax regime and of reducing the ability of the integrated oil companies to transfer their tax liabilities between upstream and downstream operations.

No such difficulties were thought to exist in the natural gas market since gas is normally sold on long-term contracts between independent parties. Invoiced prices therefore form the basis of taxation for sales of natural gas.

The norm price system is not intended to yield extra tax revenue to the state. The Act states that the norm price shall be 'equivalent to the price at which petroleum could have been sold between independent parties in a free market.'[4] Norm prices are stipulated in arrears for each quarter by the Petroleum Price Board (PPB) which consists of five members appointed by the king. Factors taken into account in establishing these prices include realized prices for open market sales of Norwegian and other crudes (with adjustments for quality, transport costs, etc.), 'official' crude prices quoted by other state oil companies and producer governments and refinery netbacks. Producer companies on the Norwegian Continental Shelf are required to provide the PPB with all the information required for norm price stipulation. They may also furnish any additional evidence that they consider to be relevant to the valuation of petroleum for the three months in question. In practice, discussions are usually held with the major producers before the stipulation of a norm price to try to avoid major dissensions. If any company considers that the stipulated prices are unreasonable they can appeal to an independent committee of 'three wise men', appointed by the Ministry of Petroleum and Energy, which has the right to decide whether or not the prices are 'clearly unreasonable'.[5]

Gross income from produced petroleum for which a norm price is stipulated is derived by:

— regarding sold petroleum as having been sold at norm price
— regarding petroleum transferred for own use as having been transferred at norm price
— valuing stocks of petroleum at norm price

In other words all transactions are deemed to take place at the norm price. Petroleum becomes taxable when it passes the 'norm price delivery point'. This point is stipulated for each field by the PPB. For example, Ekofisk is priced f.o.b. Teesside, Murchison f.o.b. Sullom Voe, and Statfjord f.o.b. loading platform. Thus for tax purposes the actual place of delivery is irrelevant; income is assessed using the delivery point prescribed in the norm price stipulation.

When petroleum is sold with a credit period greater than that assumed in the norm price (usually thirty days) the norm price for

[4] *ibid*, p89.

[5] See, for example, Syverson, J. 'Petroleum Taxation in Norway', edition of *Oil Now*, published by Den norske Creditbank, p13.

that sale is increased by 0.025 per cent for each extra day (9 per cent per annum). If, however, the credit period assumed for the norm price is thirty days and the actual credit period is shorter, there is no corresponding reduction of the norm price. If the norm price stipulates a credit period of more than thirty days and the actual sale terms are less than the stipulated period but more than thirty days, the price can be decreased by 0.025 per cent for each day by which the stipulated and the actual periods differ.

If, at the end of a three-month period, a quantity of petroleum that has passed the norm price delivery point has not been booked as sold or transferred for own use, it will be regarded as stock and valued at norm price.

In fields where the state has a carried interest agreement, the price of any sales of oil between the state and the other participants will be determined by the norm price. Norm prices are also used as a basis for valuing royalty oil.

10.8 The Effect on Investment

The Norwegian tax system has been criticised on the grounds that less profitable fields are relatively highly taxed. Marginal fields can expect to pay significant amounts of royalty and Special Tax. By comparison, the UK system differentiates more between fields of high and low profitability. The Norwegian regime can be regarded as part of an overall depletion policy which favours development of large fields rather than small ones. The only small fields currently being developed are 'satellites' of Ekofisk or Statfjord.

An interesting development is the Government's response to the Ekofisk Waterflood Project. This was proposed by Phillips in 1982 and scheduled to start in 1987. It was estimated that it would increase the overall recovery factor of the Ekofisk field by about 3 per cent. However, in June 1983 Phillips decided that the return on the project, given the risks involved, was too marginal to warrant going ahead. The Government then undertook to review the tax laws to improve the economics of the project and allow it to proceed. The adjustments subsequently made for this project were:

— Investments were to be written off 100 per cent in the first year of expenditure (instead of 16.67 per cent per annum from the year of production start-up).
— Uplift was to be deductible for only three years (instead of fifteen) at 6.67 per cent. The uplift period was to start the year after the investment had been written off.

This solution attracted considerable attention since it involved, for the first time, a tax adjustment aimed at improving the economic prospects of a major project that would otherwise have been impracticable owing to the high taxation level in Norway.

In 1983 a group known as the Øien Committee was set up to examine the fiscal structure in Norway. It did not consider any general reduction in tax levels but proposed some changes aimed at increasing the incentives to develop new fields and reduce the discrimination against new entrants.[6] The two main proposals of the Øien Committee are summarized below:

	Current	Proposal A	Proposal B
Special Tax (%)	35	35	35
Uplift (%)	6.67% over 15 years	0	0
Depreciation (%)	100	115	100
Interest deductible for ST	Yes	No	Yes
Start of depreciation	Year of production	Year of expenditure	Year of expenditure
Depreciation period (years)	6	6	6
Start of uplift	Year after production	n/a	n/a
Uplift period (years)	15	n/a	n/a
Compound rate for unused depreciation (%)	0	14	12

The first choice of the Committee was Proposal A. Key features of both proposals were the removal of uplift for Special Tax and the acceleration of the utilisation of depreciation allowances. Where the investor has insufficient income to take advantage of the allowances he is to be allowed to compound forward the unutilised part with interest. This ensures that discrimination against new entrants, if not removed, is at least reduced.

The relative advantages of the three systems depend very much

[6] See Kemp, A.G. and Rose, D., 'Communication on Energy: The Norwegian taxation debate', in *Energy Policy*, September 1984.

on the assumptions about the level of gearing of investors (since interest is not deductible under Proposal A) and the real discount rate employed. The higher the discount rate the more attractive are the two new proposals, since the effect of discounting is to increase the value of the early allowances. Overall, the incentives to develop new high-cost fields do not seem to be significantly increased by the proposals.

Oil companies are unhappy with the Øien proposals, and it is understood that for the moment the matter has been laid to rest.

PART III

THE BRENT MARKET

CHAPTER 11

THE ORIGINS AND GROWTH OF THE BRENT MARKET

11.1 Introduction

It is not possible to date the origins of the Brent market precisely. In 1981 the market did not exist: Brent blend was simply one of several grades of crude oil available to a European or US refiner on the North Sea spot crude market. Today it has become an institution. Brent blend is now one of the most actively traded crudes in the world, a key international marker, replacing Arabian Light in the Atlantic Basin, and a highly standardized paper commodity traded on a unique informal forward market.

As far as can be seen, the development of the market was entirely unplanned. It evolved in response to changing market conditions in the world spot crude market and specific institutional factors at work in the North Sea. These included: the general expansion of spot crude trading following the decline of term contracting starting in 1979, the pressure for short selling in a falling market, the rapid growth of North Sea production, the structure of the UKCS fiscal system, and the role of BNOC in setting prices and disposing of the very substantial volumes of UKCS production available to it through participation agreements. Each acted in varying degrees as a spur to the growth of spot trading in the North Sea, sometimes independently and sometimes in concert. Yet, of these factors, tax spinning by some of the integrated oil companies deserves special consideration because it played a major role.

The market itself appears to have evolved in a number of stages, moving from one stage to the next as surges in activity and changing market circumstances prompted new styles of trading. In general, formalization lagged behind the trading reality.

The *first stage* involved short selling by traders once spot market prices had fallen below the higher-tier North Sea term prices in early 1981. This was possible because refiners were still prepared to pay for security of supply, and North Sea producers were keen to establish a market price for North Sea oil.

The *second stage* involved the introduction of daisy chains linking producers to refiners via traders. At this point chains were wet, that is to say they were multiple transactions involving a specific cargo of oil. The chains emerged as a result of a higher level of spot market activity and fluctuating expectations about the future price of oil in 1Q82.

The *third stage* involved a transition from wet chains to dry chains as the level of speculative activity increased with greater uncertainty over future prices in 4Q82. By now many traders were not concerned with physical supply of oil, but were engaged in an elaborate game of trying to guess the outcome of the OPEC meeting and the winter weather.

The *fourth stage*, and so far the final one, involved progressive formalization of trading procedures to cope with the problems of resolving an increasing number of paper deals. This process began in January 1983 with an excess of sales over actual availabilities, and has continued ever since. It included the emergence of Brent as a standardized trading commodity, rules for ending each month's business and the establishment of more formal trading hours.

11.2   The Beginnings. Short Selling by Traders in 1981

Spot trading of crude oil grew substantially following the Iranian crisis of 1978–9. At that time prices were rising rapidly and a number of traditional term supply arrangements had been broken by *force majeure*. Traders were therefore able to make money by acquiring oil from producers, holding it, and then selling to refiners. Thus the prevailing tactic was to go long. By 1981 prices had begun to fall and profitable trading opportunities had dried up as the more hawkish producers priced themselves out of the market. Traders therefore looked for new sources of oil and new trading tactics.

The North Sea was well suited to meet their requirements for a number of reasons. First, the quality and location of North Sea oil matched market requirements. Secondly, the fact that the oil was produced by the companies themselves removed concerns over security of supply, which was still a major issue at the time. And thirdly, the UKCS fiscal system created an incentive for integrated companies to trade their equity oil in certain circumstances (see Chapter 8). Such a combination of security and liquidity in a falling market suited the traders very well and the centre of gravity of spot crude trading began to shift towards the North Sea as traders started to sell short in the expectation of further price reductions.

The general willingness of traders to sell short was first noted by

Argus as early as January 1981,[1] and reports of short selling and covering by traders who had sold short were made throughout the year. In many cases short selling also meant forward selling since traders were often trying to balance rather limited North Sea f.o.b. availabilities against c.i.f. arrivals of long-haul crudes from the Middle East. Short selling was feasible because a number of integrated oil companies were now prepared to sell their equity production into the spot market for tax purposes.

This option had always been open to the companies, within the terms of the 1975 Oil Taxation Act, but had not so far been used since spot market prices had generally been close to or above the BNOC term price. The position was reversed in January 1981 as spot prices fell below the relatively high BNOC price and several integrated producers were quick to take advantage of the opportunity. At this time the intention was almost certainly to put pressure on BNOC to reduce the term price to a more moderate level (viz. equivalent to Saudi Arabia) rather than to sell all equity production on the spot market. If so, it certainly succeeded.

Tax spinning as it has now become known was first reported by *PIW* in April,[2] and is thought to have been initiated by Conoco, closely followed by BP. At any rate BP is reported to have sold Forties to the US Strategic Petroleum Reserve at $1.75 below the BNOC price in May 1981[3] and spot selling by integrated companies was widely reported in June 1981.[4] During this period the most commonly traded crude was a Brent/Ninian/Forties option usually one, but occasionally two, months ahead.[5]

11.3   The Emergence of Wet Chains in 1982

In early 1982 the North Sea market moved into a second stage of development. The spot price collapse in February and March (Brent fell from around $36 in January to below $28 at the bottom in March) encouraged short selling as price expectations plunged. Liquidity also rose as many majors with equity production sold forward to minimize losses (both on the trading front and the tax front). As a result the trading horizon moved further forward with a

[1] *Petroleum Argus*, 30th January, 1981
[2] *PIW*, April 20, 1981
[3] *Petroleum Argus*, 8th May, 1981
[4] *ibid*, 5th June, 1981
[5] *ibid*, 5th June, 10th July, and 25th September 1981

number of May barrels being done in early March.[6] However, once spot prices began to pick up again in April the amount of physical oil available to meet short sales fell sharply and cargoes were passed from hand to hand creating an early version of the now famous daisy chains. At the end of April *Argus* reported: 'A May cargo of Brent, not recently traded, has already passed through 18 companies' hands, including four majors and some duplications.'[7] The general firming of the market continued through the summer, and there appear to have been fewer spot sales by the majors. Nonetheless short selling by traders continued and *Argus* noted in August that 'circumstances such as the need to cover short sales can bring about a temporary hardening for certain crudes', especially Brent.[8]

11.4   The Transition to Dry Chains in 1982–3

The transition from wet chains to dry chains appears to have taken place in the autumn of 1982. First, an unexpected reversal of spot and forward prices in September (a contango in which prompt Brent fell 20–25 cents below forward) caught a number of short traders out leaving some with dry barrels according to *Argus*.[9] Then, in November and December, as forward prices for January Brent began to fall and traders tried to guess the outcome of the next OPEC meeting and bet against winter weather, the level of speculative trading increased substantially.

In late November *Argus* reported: 'Short sellers have been active on January Brent and some have already covered from other short sellers'[10] and in early December that it had been a 'week characterised by vigorous trading in paper barrels of January Brent'.[11] This theme was echoed by *PIW* who asked 'whether recent sharp fluctuations of spot oil prices are really a useful indicator of trends or just pure speculation'.[12]

Once January 1983 became the operational month problems began to arise. *Argus* reported: 'long chains of traders involved in January Brent [have] led to problems on timing there. Several of

---

[6] *ibid*, 12th and 19th March, 1982
[7] *ibid*, 30th April, 1982
[8] *ibid*, 20th August, 1982
[9] *ibid*, 10th and 15th September, 1982
[10] *ibid*, 26th November, 1982
[11] *ibid*, 3rd December, 1982
[12] *PIW*, December 13, 1982

these January Brent cargoes are not yet covered which has encouraged holders of wet barrels to hold off.'[13] *PIW* said: 'The volume of North Sea crude oil actually changing hands in spot deals between producers and refiners has been dwarfed by an avalanche of paper transactions among speculators – but this time the house of cards seems to be on very shaky ground. More cargoes of Brent crude have apparently been sold than there is oil available for loading in late January, but many of the sales are likely speculative and not really intended for end-users.'[14] According to *Argus*, confusion reigned through the remainder of January as those involved tried 'to finalise or otherwise settle the complexity of January Brent deals'[15] and *PIW* reported: 'Several primary suppliers of North Sea crude are working to clear up the logjam, and some are refusing to load cargoes unless the ultimate real end-users are identified.'[16] In the end matters were resolved and February appears to have worked smoothly. *Argus* recorded in early February: 'Most short sales of Brent for February appear to have been covered and interest is turning now to March barrels.'[17]

In retrospect this episode is interesting because it provides a number of insights into the development of the Brent market:

— First, it demonstrates the scale of speculative or paper trading that was taking place in the North Sea market as early as 4Q82.
— Secondly, it points to the informal structure of the Brent market which was initially unable to cope with a sudden increase in a relatively new form of trading.
— Thirdly, and most significantly, it reveals the extent of the support of the major integrated oil companies for the new trading patterns which had evolved. Without the goodwill of those companies, which had primary access to wet barrels in the North Sea, the entire speculative structure could have collapsed in January 1983, leaving many traders with contracts they could not fulfil. It would, of course, be wrong to suggest that such goodwill was purely altruistic. Commercial interests alone would have given primary suppliers an incentive to help resolve problems. If the majors had been unhappy with the system they could certainly have demolished it, but they needed it, not least for the perceived advantages of tax spinning.

[13] *Petroleum Argus*, 7th January, 1983
[14] *PIW*, January 17, 1983
[15] *Petroleum Argus*, 19th January, 1983
[16] *PIW*, January 17, 1983
[17] *Petroleum Argus*, 9th February, 1983

— Finally it provided an incentive for those involved to institute more formal procedures to help resolve the elaborate structure of paper trading.

11.5   The Formalization of the Brent Market

From January 1983 the market moved into its fourth phase of development as the level of activity continued to increase and efforts were made to introduce formal rules to ease trading difficulties. At this stage Brent/Ninian/Forties options were still being traded, although these were gradually phased out over the year as activity focused increasingly on Brent in response to pressures for a more standard trading commodity.

Amongst the majors, BP seems to have been the most active in pressing for greater standardization of trading procedures. Further pressure also came from the opening of a formal crude oil futures contract on the New York Mercantile Exchange at the end of March 1983. This created opportunities for arbitrage and gave traders greater familiarity with futures market procedures. Progress appears to have been slow however. On the one hand *Argus* reported in March that a 'burst in activity following the OPEC meeting' included 'several bookouts'.[18] On the other hand June availabilities were withheld until late in May[19] and availability became so tight that 'one trader was forced to take Forties.'[20]

In November 1983, the London oil futures market, the International Petroleum Exchange (IPE) tried to launch a formal Brent futures contract in emulation of the successful Nymex WTI contract. However it never really took off and was withdrawn by March 1984. At the time its failure was ascribed to restrictive delivery terms (1000 barrel lots, in tank ARA), which did not match trading reality. In practice it probably failed because there was no incentive for Brent paper traders to move from their exisiting unregulated and clearly profitable market to a regulated and formal exchange. In November 1985 the IPE launched another Brent futures contract which is based on cash settlement and therefore avoids the problems associated with physical delivery.[21]

During 1984 most of the rules that now characterize the Brent market were introduced: the nomination procedure at Sullom Voe

[18] *ibid*, 18th March 1983
[19] *ibid*, 20th May, 1983
[20] *ibid*, 17th June, 1983
[21] *ibid*, 18th October, 1985

and the pattern of trading within the month; the obligation to give fifteen days' clear notice of the first day of a three-day loading range; a formal end to business at 1700 hours London time as far as nominations are concerned; the legal and procedural background to book-outs; the use of telephones rather than telexes; and the need for letters of credit. All these of course are only useful as long as the majority accept them, although the majors appear to some extent to act as policemen ensuring that business proceeds smoothly.

Today the market seems poised for further change. One major player and source of liquidity, BNOC, has been abolished. It will be replaced by a smaller royalty trading agency which will give increased availabilities to some medium-sized North Sea producers, possibly bringing more players into the market. The spinning debate continues as the Oil Taxation Office has yet to make any clearer the rules for taxing internal transfers of North Sea production within the integrated companies. More end-users are thought to have entered the market. Brokers, some of whom specialize in Brent, now play a more active role: creating opportunities, testing market limits, and generally trying to increase the level of activity. The IPE is attempting to bring non-oil speculators into the market by means of its new Brent futures contract and Chase Manhattan has launched a pilot documentation scheme for Ekofisk loadings in a bid to set up a clearing house for Brent market transactions. Opinion is, however, divided as to whether the market will continue to grow or start to shrink and whether greater formalization and control are desirable.

CHAPTER 12

THE MODE OF OPERATION OF THE BRENT MARKET

12.1 Introduction

The market commonly known today as the Brent market is a relatively new development in crude oil trading. A few years ago it was not distinguishable from other spot crude markets either in the North Sea or elsewhere; but it has recently evolved into an exceptionally active and highly specialized, though informal, institution which has become one of several barometers of world oil prices.

The Brent market is a curious hybrid of spot and forward transactions which allows companies to combine their own traditional oil trading techniques with those more usually associated with other primary commodities. In its present form it is an informal, self-regulating club of North Sea producers, oil traders, refiners and brokers, each of which is in the market for a variety of different reasons.

Such rules as exist either derive from the operational conditions governing the production and shipment of Brent blend, or have been created to overcome specific problems that have arisen in the functioning of the market during its short existence. These rules are, of course, not immutable and are only effective as long as the majority of participants accept them. In practice circumstances change and so, therefore, do the rules. So far attempts to achieve greater formalization of the market have been resisted by a significant proportion of the participants. Any description of the market and its operation is thus best regarded as a snapshot.

12.2 The Mechanics of Trading

We shall describe in this section:

— the nature of the commodity traded in the Brent market
— the type of transactions
— the method of doing business
— the organization of the market

(a) The Commodity. Brent blend is a light, low-sulphur crude oil, typically around 38°API gravity. It is acceptable to most refiners in the Atlantic Basin for most purposes, including lube feedstock manufacture. In practice the precise quality of the mixture will vary with the production of the component fields: Brent, Cormorant, Deveron, Dunlin, North West Hutton, Hutton, Murchison and Thistle. No explicit price adjustment is made since fluctuations in quality are regarded as a buyer's risk, but major shut-downs are usually notified in advance and the market valuation will change to compensate.

Brent blend is collected by a pipeline system which discharges into tank at Sullom Voe on Shetland. It is available f.o.b. the Brent terminal in minimum parcels of 600,000 barrels (or 80,000 tonnes). Until the end of 1984, the minimum parcel was 500,000 barrels but higher production of both Brent and Ninian (also available at Sullom Voe) began to create congestion at Sullom Voe and so the parcel size was increased. For this reason loading schedules will sometimes also be rearranged to accommodate VLCC liftings. In January 1985, with Brent blend production running at just over 1 mb/d, there could theoretically have been fifty-two Brent parcels available from the month's production (prior to 1st January there could have been sixty-two); in fact only forty vessels loaded since there were several VLCC parcels.

(b) Types of Transaction. The Brent market involves two main types of transactions:

— Dated cargoes sometimes referred to as dated Brent
— 15-day cargoes also called 15-day Brent

A dated Brent deal is the sale/purchase of a specific cargo of oil made available within a determinate date range. It is a normal spot market transaction similar to others elsewhere in the North Sea or in the rest of the world.

A 15-day Brent deal is something quite different. It refers to the sale/purchase of a cargo for delivery on an unspecified day of a given month. The actual delivery dates are eventually determined by the seller with a minimum notice of fifteen calendar days.

It is useful to relate the concepts of dated and 15-day cargoes to other pairs of terms relevant to an understanding of the market. The relevant pairs are:

— spot and forward
— wet barrels and dry or paper barrels

To all intents and purposes a dated cargo is a spot transaction. In theory the specified delivery dates of a dated deal can be in a different month from the one in which the contract is made, in which case the dated cargo is a forward transaction. In practice this does not usually occur because sellers cannot determine the precise date of their availabilities in future months.

A 15-day Brent deal is a forward transaction made at least sixteen days before the first possible delivery date. In the first half of, say, August, any participant can enter into a 15-day deal for August, September or October (etc.) Brent; but in the second half of August, August Brent will cease to be available in the 15-day market and the deal may only involve September, October or November (etc.) Brent. Thus a 15-day deal for a particular month can sometimes take place in the same month, but many such 15-day deals take place in earlier months.

A wet deal (wet barrels, wet Brent) refers to a physical transaction in which a specific cargo actually changes hands. Thus a dated deal is usually a wet deal. A dry or paper transaction (paper barrels, paper Brent or dry barrels) refers to the sale/purchase of a claim on a cargo of Brent. A 15-day Brent deal ultimately refers to a wet transaction (unless a book-out takes place, see 12.2(e) below), but because of its forward character it can and usually does give rise to intervening sequences of paper transactions.

Dated and 15-day Brent are, of course, interchangeable (subject to timing), since they both involve the same commodity: the Brent system blend of crudes. However, the purpose and method of trading is different for each, and for this reason they may be considered analytically either as two different commodities or as a single commodity traded under different sets of rules in two distinct markets.

(c) Sales Procedures. The procedures governing the loading pro-gramme at the Brent terminal at Sullom Voe explain certain features of 15-day Brent deals.

First, the earliest date on which the actual day-range (a three-day window) for delivery of a specific cargo in a particular month can be given is the 15th of the preceding month. This is because the loading programme, organized on behalf of the Brent system participants by Shell UK (the terminal itself is operated by BP, and the port by the Shetland Islands Council) involves the following procedures:

— Producing companies (and BNOC) nominate their preferred loading dates for the relevant month by the 5th of the preceding month.

— Shell then confirms the whole month's programme by the 15th of the preceding month.

Secondly, a company wishing to trade further forward is in no position to specify the actual delivery day-range because the loading programme is not yet determined. It can only sell for that forward month. For example, a company with Brent availability willing to enter into a forward deal on 10th August can sell September or October (etc.) Brent but not a specific cargo dated 10th–12th September or 5th–7th October (etc.). The procedures governing the loading programme determine the institutional features of oil supplies to the Brent market. They explain the terminology in use (e.g. 15-day Brent), and, more fundamentally, the important distinction between the two commodities (dated and 15-day) and the differences in the modes of transaction.

Both dated and 15-day Brent are sold f.o.b. on identical terms: payment is due thirty days after the date of the bill of lading; insurance, freight, and ocean losses are the responsibility of the buyer; demurrage at the terminal is the responsibility of the seller. However, dated Brent is sold as a specific cargo available in an actual three-day loading range. Once a cargo has loaded it can theoretically be sold or exchanged, but for most European destinations the sailing time is too short to create worthwhile trading opportunities. US destinations on the other hand allow sufficient time and opportunity for trading and cargoes often change hands on arrival in the USA.

In contrast 15-day Brent is bought and sold up to three (or occasionally four) months ahead of the date of loading. In this case oil is sold 'for the month'. The delivery dates are at the seller's discretion subject to the conditions mentioned before (notice, loading date range, etc.). The key difference between 15-day and dated Brent is that a 15-day cargo can be sold without physical cover in the first instance. As a result participants can take short or long positions in the market, buying or selling cargoes for forward months. 15-day Brent has therefore become, in some respects, like a futures commodity with the bulk of the trading activity taking place in paper barrels since many of those involved do not wish to take physical delivery of the ultimate parcel of oil. Despite the emphasis on paper trading, most sellers take the obligation to supply physical oil seriously as they cannot predict in advance whether or not a particular buyer will insist on physical delivery.

(d) The Methods of Doing Business on the Brent Market. The Brent market is not organized like a formal commodity futures market

although it exhibits a few similar characteristics. Forward Brent shares with futures markets the following: both involve highly standardized contracts which can be used for speculation, hedging and arbitrage with other oil (and indeed with other commodity) markets; and, as in any commodity futures market, the physical characteristics of 15-day Brent deals are all specified and uniform.

There are, however, important differences between forward Brent and futures markets. These are as follows:

— The Brent market is not formally institutionalized and regulated: it is more like a club than an exchange. In this context, the important difference is that bidding does not take place on a floor in a way that ensures full transparency of prices and volumes. Contracts are negotiated directly between participants and nobody is required to say whether a deal has been entered upon or to disclose the terms of the transaction.

— The standard parcel in Brent forward dealings is large (500,000 barrels until 1985 when it was set at 600,000 barrels) compared with the Nymex parcel (only 1000 barrels). This feature tends to restrict entry as few individuals and not all companies can raise the credit required or take the potential loss on a single deal.

— The Brent market, though it involves long chains of paper deals, is strongly rooted in physical trading: a significant proportion of the Brent blend output is physically disposed of in that market. In a futures market such as Nymex, the physical transaction is rarely the motivation for or the objective of trading. It can be construed in many instances as a sanction for failing to liquidate paper contracts before the maturity date.

The Brent market has evolved its own set of rules and methods of clearing the complex relationships between buyers and sellers that emerge during a trading month. Since there are no institutional procedures for membership, there is no criterion for participation other than actual dealing. But the ability to enter an actual deal largely depends on the identity of the firm and its reputation. The informal character of the market and the fact that the standardized transaction is a very large parcel of oil together impose more severe restrictions on entry than the statutes and rules of a formal futures market.

All business in the Brent market is conducted on the telephone, either directly between participants or through a broker. The majority of the participants know each other and are aware of their respective track records. New entrants are normally introduced by a broker and operate through him until they have acquired a sufficient reputation. Since it is a fast moving market with a high turnover all deals are made verbally; telex confirmation then follows. In principle, a paper Brent deal can be executed very quickly: the commodity is standardized and so the only point to negotiate for a given month is price. However, since the rules remain informal, a careful trader will check each point thoroughly before committing himself with the magic words: 'I will do you a Brent at $X per barrel.' Not all traders are equally careful and there have been cases where both partners thought they were buying!

On the surface the costs of a Brent transaction appear to be low: simply a telephone call and a telex confirmation. In practice they are much higher. To begin with, the market is not transparent: up-to-date information about current trading levels is not publicly available and can only be obtained by constantly sampling the other participants. Published price data are at best several hours old and at worst wrong. An active presence in the Brent market therefore requires dedicated manpower. This information bottleneck has created a niche for brokers who not only provide information to active traders but can also be the only safe way of doing business for a minor trader. The current broking fee is 2 cents per barrel, although there are moves to reduce this to 1 cent per barrel.

Many sellers require buyers to provide a letter of credit from a bank: this can cost between 1 and 4 cents per barrel depending on the party involved. Major companies, especially those with Brent availabilities, such as Shell, Esso, BP or BNOC, are not asked to supply letters of credit, which of course gives them a trading advantage. Finally the administration costs of following up each transaction, checking letters of credit, identifying chains of entitle-ment, passing on the nomination, bills of lading and so on all take time and involve further costs.

(e) The Clearing of the Market. In the Brent market, there is no exchange or regulatory body to match sellers and buyers; there is no end-of-day closing when the various short and long positions are matched up and the open commitment established for all forward months. Therefore the process of rationalization involves all the participants as there are usually more commitments than wet cargoes.

Since Brent deals are concluded by direct agreement between the two partners concerned all that exists initially is a series of pair-wise deals. Take for example the set of transactions made for a given month, say August Brent. These could have been agreed at different times during the preceding four, or perhaps five, months. As participants move to cover their positions, or go longer or shorter for August Brent, a network of potential chains of commitment begins to emerge. In the rather poetic terminology of the Brent trade a sequence of commitments (or of paper deals) is referred to as a daisy chain. The point to remember here is that the daisies are not pre-arranged in chains of given length with determinate sequences so long as trading for our illustrative August Brent goes on. Alternative sequences and chains of different length may be made with the many daisies scattered all over the market.

The daisies are sewn in an actual chain when a primary seller serves a 15-day notice of delivery (in our example of August Brent, on any day between 16th July and 13th August) to one of his buyers. This first buyer, depending on whether or not he is an end-user willing to lift the oil at the specified date and on whether he holds sales contracts for August Brent, may serve the same notice on a company that has bought an August cargo from him. The process continues until somebody takes the oil, either because he has some use for it, or because he has no sales contract for August Brent in his portfolio, or because he is left with insufficient time to give fifteen clear days' notice to the next participant down the line.

Thus the daisy chains only appear when primary sellers begin to dispose of the physical output of a month (say, August Brent) in a market that has collected paper claims for that month's production. In any particular month there are about 40–50 Brent cargoes to be lifted according to a loading programme determined by the producers and a number of market participants holding one or several purchase contracts matched or unmatched in their portfolio by sales contracts. The daisy chain links the primary seller who has announced an intention to deliver with one of the many holders of a paper contract for a cargo. The participant who ends up with the cargo is identified by a process which resembles the children's game of 'pass the parcel'.

As mentioned before, the actual path of a Brent daisy chain is not predetermined. It depends on the sequence of nominations that takes place once the month has become wet, and will usually involve a choice by those participants who hold more than one paper Brent position. As each participant is obliged to give a minimum of fifteen days' notice, the length and course of chains can depend on the time available to pass on the nomination. Once the fifteen days have expired, nobody is obliged to accept the nomination, and a company that thinks it has covered itself by selling paper Brent could find itself not only taking delivery of dated Brent but also obliged to find a further cargo of Brent for a date at least fifteen days ahead! In some cases a nomination will be accepted on a shorter notice subject to the agreement of all further parties involved. Thus the rules of the nomination procedure can create further uncertainty and risks for those involved in the Brent market.

Market participants sometimes succeed in cancelling their obligations to buy/sell, say, August Brent by agreeing among themselves on a book-out, before the date on which delivery notices would be served (i.e. before 15th July). The possibility of a book-out arises when a potential daisy chain involves the same participant more than once. In these circumstances there is a potential 'loop' in the chain which can be cut off the chain if all participants in the loop agree to give up their rights to acquire a cargo and their obligations to provide it by discharging these respective commitments in paper.

Normally a book-out is initiated by a company willing to liquidate a position. Such a company will start by identifying a potential chain with a relevant loop, and will then attempt to persuade others in the loop to cancel their mutual obligations.

The book-out takes place on the basis of an agreed reference price, with a cash settlement for the difference between this reference price and the prices in the contracts put for cancellation. Book-outs are normally settled in advance (by the 15th of the preceding month) to avoid the risk of default by a member in the relevant loop. In case of failure the ultimate commitment to buy/supply physical oil is re-established.

Not all participants are prepared to agree to book-outs since some are primarily interested in obtaining physical Brent (wet barrels) at the end of the day. Other companies have tended to abuse the process by only agreeing to book-outs where they have made a profit and can therefore realize cash up front; this leaves the loss-making deals to be settled in the wet market. One company, BP, appears to have been instrumental in organizing the book-out procedure, writing the standard telex and circulating a proposed set of procedures. In a small number of special circumstances two companies holding complementary positions (buy-sell/sell-buy) will reach a 'cancellation agreement' which removes that pair of transactions from the market.

When the month (say, August) becomes wet, that is to say when the loading programme for August Brent is agreed and primary sellers begin to serve 15-day notices (which they can do on any trading day between 16th July and 13th August), the market for August Brent will have to clear. Participants who have booked out or cancelled their contracts are out of the game, but the sewing of daisy chains will actualize a number of wet deals between primary sellers and ultimate buyers.

The clearing process is complex, not only because the nomination procedures may give insufficient time to establish the most efficient patterns of daisy chains, but also for the following reasons:

— All paper contracts involve a binding commitment to buy (or sell) physical Brent if the participant is called upon to do so. To avoid this commitment the contract has to be cancelled through an explicit book-out agreement. It is not sufficient to hold a balanced portfolio with an equal number of sales and purchases contracts, because any one of these contracts could become wet in its own right.

— The number of wet cargoes that become available in the operative month need not correspond to the sum of net positions established for that month. This is because any participant in the market can initiate a 'sell' of Brent whether or not he is a primary seller with physical availabilities, and whether or not he has contracted to buy an amount of Brent equivalent to his sales commitments.

It should be recalled that companies operating on the Brent market often wish to cover their positions. In the period up to the 15th of the preceding month (July in the case of August Brent) this can only be done in paper transactions, but such a cover is not necessarily secure. As mentioned before, any paper contract in a balanced portfolio may become active at some point in time and destabilize the position if the holder fails to activate the matching paper. The failure to match actual deals, when the paper portfolio is balanced, arises because the exact maturity of a Brent contract for a given month is indeterminate within a 30-day period, or because the time available to pass on the nomination may run out. In our example, the maturity date of August Brent for a particular contract will fall on any day between the 1st and the 31st of August, and will become known to the participant some time between mid-July and mid-August.

Once the relevant month becomes wet, a company wishing to cover its position has a choice between covering in paper Brent (up until the 15th of the relevant month) or covering with a dated cargo. It is at this point therefore that the paper market and the physical market come together. A company without physical availabilities will need to purchase a suitable dated cargo from someone with an availability if it cannot cover in paper.

At this point the market becomes unbalanced, and there can be a scramble for dated cargoes to meet paper obligations as the number of trading days for the relevant (Brent) month diminishes. Should such a situation arise, those companies with physical availabilities at Sullom Voe can extract a substantial premium for wet barrels.

Other clearing mechanisms can also be tried. Buyers may be persuaded to accept a substitute crude at a favourable price, or they may be prepared to renounce their rights for a consideration. The reverse effects obtain when the market is undersold.

To sum up; the forward market for a given (Brent) month is cleared through a number of processes:

— book-out and cancellation before contracts mature
— the sewing of daisy chains which link the primary seller of wet barrels with the holder of a purchase claim willing to take delivery (or obliged to accept delivery because he is unable to pass the nomination on to another participant)
— the pricing mechanism which pushes the price of dated cargoes up or down according to the relationship between net positions in paper and available cargoes
— recourse to substitute crudes for settling contracts in case of shortage

12.3   Further Remarks on Daisy Chains

It may be useful for the purposes of clarification and illustration to analyse the evolution of an actual daisy chain. This analysis refers to a chain presented by Silvan Robinson of Shell in a talk at the sixth Oxford Energy Seminar in September 1984. It was subsequently reported by *PIW*.[1] The ultimate chain, which is shown below,

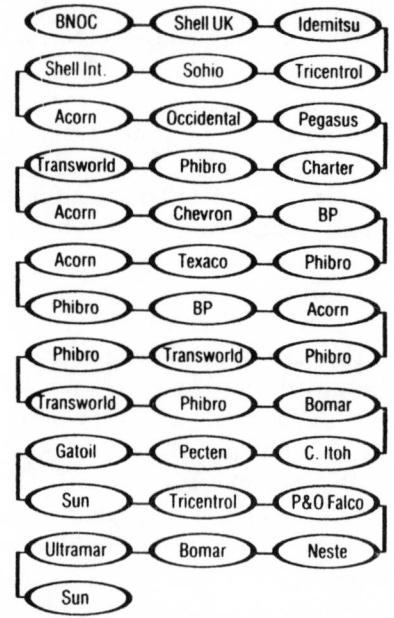

[1] *PIW*, November 12, 1984

involved thirty-six separate transactions for March Brent by twenty-four different participants which took place in the first quarter of 1984. We have been able to establish that the physical cargo was nominated by Shell rather than BNOC (Brent parcel number: B0333) for the 19th–21st of March and was lifted from Sullom Voe by Sun on 21st March 1984 on the *Viking Eagle* and delivered to Sun's Marcus Hook refinery in the United States. However, the chain is presented here as published initially. The chain itself is rather longer than was usual for this period, and the composition of the chain suggests that it would have benefited from a number of book-outs. Without book-outs the nomination procedure must have been very slow since there were thirty-five transactions between BNOC and Sun.

A Brent daisy chain evolves in three stages. First, paper or 15-day Brent is bought and sold by market participants for a specified forward trading month. In this case the majority of the transactions appear to have taken place in January and February, with February being the more active month:

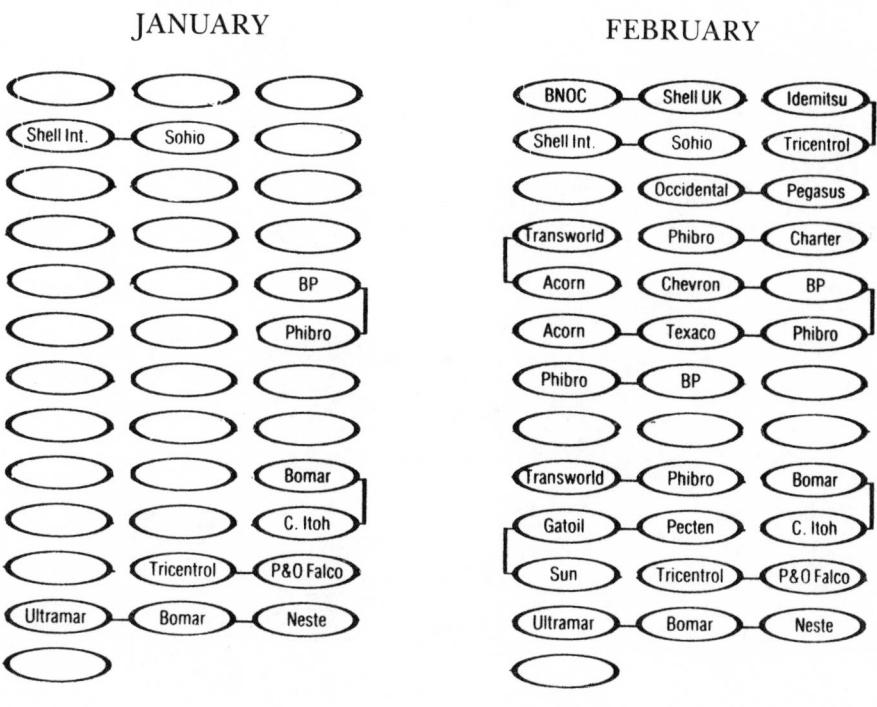

JANUARY   FEBRUARY

In both months the commodity being traded is 15-day March Brent. At this point no actual chain exists, only a number of pair-wise deals between participants who can potentially link up to form a chain, either as part of a book-out or once a cargo has been nominated by a primary supplier. As a result the final order of the daisy chain need bear no relation to the date order of the transactions.

Secondly, the more active companies (especially those not interested in taking physical delivery of oil) try to identify possible chains of commitment linking a 'sell' and a 'buy' position. If the other companies involved agree, a book-out is arranged, which effectively nets out a sequence of transactions from the chain. In this case, as mentioned in section 12.2(e), the companies involved agree a reference price and pay each other the difference; this is usually done in advance to avoid the risk of default once the trading month becomes operational (or wet). Several possibilities for book-outs which could have taken place in the Silvan Robinson chain can be identified. Any of these could have taken place. The fact that a

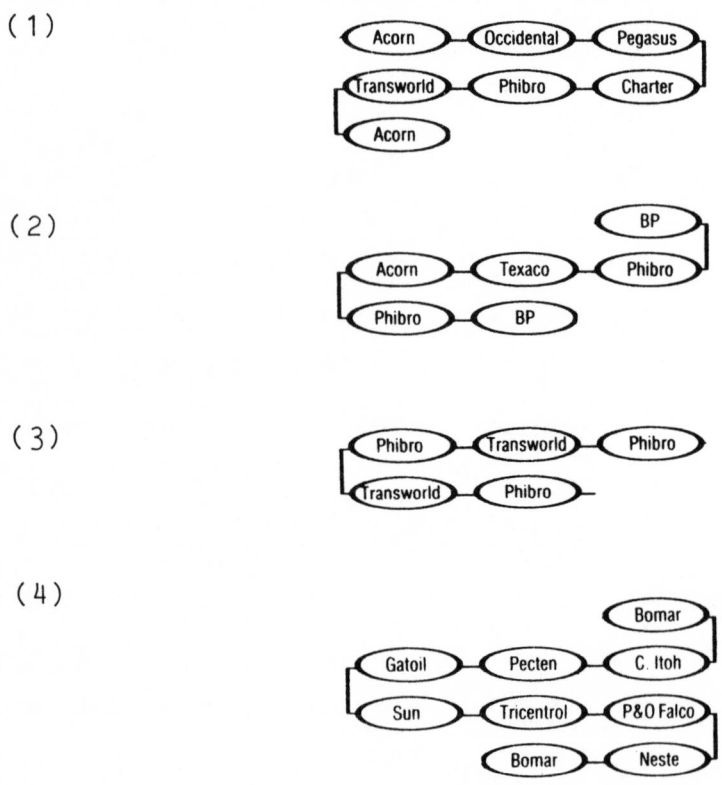

number of companies occur more than once (Phibro, BP, Tricentrol, Bomar, Transworld, Acorn – a Charter subsidiary) creates many possibilities for book-outs. The most effective book-out would have linked the two Tricentrol deals, removing twenty-eight transactions from the market prior to nomination.

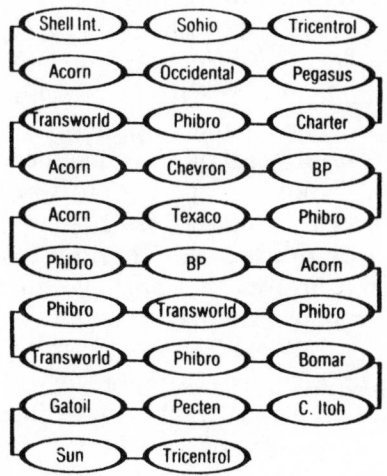

The third stage involves the final nomination of a wet cargo by a primary seller, in this case Shell UK for 19th–21st March. At this point the nomination is passed from hand to hand as long as the companies involved can give fifteen days' clear notice of the first day of the loading date range. The chain ends either when the fifteen days expire or when it reaches a company wanting to lift wet barrels, in this case Sun. In our example the final transaction apparently took place on 9th March, some time after the last moment at which fifteen days notice could have been given (i.e. 1700 hours on 3rd March). This is possible as long as the remaining participants in the chain agree. It is during the third stage that the daisy chain actually crystallizes from the range of possible chains that might exist prior to nomination.

12.4 Conclusions

The Brent market is both spot and forward. Spot transactions involve actual cargoes available for prompt or precisely dated deliveries. Forward transactions are for the Brent output of a given month. Any market for forward physical transactions gives rise to the trading of paper claims. This is a natural development which is initially due to the existence of a time interval between the day a deal is made and the day on which it matures. Participants can only initiate sales that they hope to cover at some future date during this time interval, in which some buyers have the opportunity to change their minds and to sell the contract to somebody else. In short, a paper market is the product of forwardness even when the original intention was to deal forward in physical barrels.

Once this development occurs the paper market takes on a life and acquires a momentum of its own. This is precisely what happened with Brent. The volume of paper transactions will tend to exceed the amounts of physical oil to which these paper deals relate by a large (but variable) factor. The forward market provides opportunities for speculation, and, because of its physical character, it can be used by producers and end-users for hedging. The presence of hedgers ensures that speculation is a non-zero-sum game, even when the market has reached some form of stable behaviour.

The most interesting characteristics of the forward Brent market are that the commodity traded is the Brent blend output of each month, and that the maturity of any contract is left indeterminate within a 30-day range. This feature restricts the scope for true hedging on the part of end-users because the delivery date is not known at the time of dealing and thus an important uncertainty remains. This feature of the Brent market also leads to a second round of trading when the relevant month becomes wet (from the 15th of the preceding month to the 15th of the relevant month), partly because the forward market for that peculiar commodity – the output of a month – has to be cleared over these thirty days and partly because the existence of wet availabilities that can be delivered prompt creates opportunities for spot deals. Price movements become exceedingly difficult to interpret since they are the result of different forms of trading undertaken simultaneously for different reasons and interacting with each other.

Finally, the fact that the Brent market is not institutionalized like an exchange raises interesting questions about its vulnerability to accidents and its efficiency. A market can be hit by a crisis of confidence when a series of defaults occurs or when it is affected by a speculative fever. The mode of operation of the Brent market provides protection against serious default because entry is restricted by the size of parcels ($15–18 million), which keeps out participants of meagre financial means, and by the virtual impossibility for an entrant with insufficient credentials to find a partner for a deal. Further, some major oil companies informally exercise the role of supplier of last resort. They watch the market and will ultimately provide wet barrels at a premium to the participant who fails to clear his position. Of course the system is not absolutely secure but it has a number of built-in safeguards.

The efficiency issue is more complex. The absence of complete transparency gives a strong advantage to big participants with a wide network of contacts in the market and a large team of traders continually engaged in monitoring activity and initiating deals. The fact that forward contracts do not mature on a given day but within a 30-day period (however justified on logistic and technical grounds) may not be very efficient from an economic point of view. In this system prices do not provide very clear signals about underlying demand and supply trends because they also tend to reflect complex adjustments (taking place and overlapping with each other during the relevant month) to both ephemeral and fundamental distortions. Yet Brent prices play a barometric role for other parts of the world petroleum market, and though they may provide useful indications of the true state of play, they can also cause unnecessary commotions.

APPENDIX TO CHAPTER 12

TYPICAL BOOK-OUT TELEX

From Company A
To Company B (UK)
   Company C (NY)
   Company D (Houston)
   Company E (UK)

It has come to our attention that the following sequence of agreements exists for a cargo of Brent system crude oil which is to be delivered during June 1985.

Company A proposes the following cancellation agreement format in order to facilitate the handling of these agreements.

The language of the cancellation agreement would read as follows:–

Quote

This agreement is dated as of (finalisation date) by and among company A and company B and company C and company D and company E

Whereas (a) companies A and B are parties to a contract dated 13th March 1985 in which A has agreed to sell, and B has agreed to purchase a cargo of Brent system crude oil. Such contract as briefly described as follows:–

Volume	Crude Type	Delivery
600,000 barrels	Brent System	June 1985

and whereas (b) companies B and C are parties to a contract dated (please advise) in which company B has agreed to sell, and C has agreed to purchase a cargo of Brent system crude oil. Such contract as briefly described as follows:–

Volume	Crude Type	Delivery
600,000 barrels	Brent System	June 1985

and whereas (c) companies C and D are parties to a contract dated (please advise) in which C has agreed to sell, and D has agreed to purchase a cargo of Brent system crude oil. Such contract as briefly described as follows:–

Volume	Crude Type	Delivery
600,000 barrels	Brent System	June 1985

and whereas (d) companies D and E are parties to a contract dated (please advise) in which D has agreed to sell, and E has agreed to purchase a cargo of Brent system crude oil. Such contract as briefly described as follows:–

Volume	Crude Type	Delivery
600,000 barrels	Brent System	June 1985

and whereas (e) companies E and A are parties to a contract dated 8th May 1985 in which E has agreed to sell, and A has agreed to purchase a cargo of Brent system crude oil. Such contract as briefly described as follows:–

Volume	Crude Type	Delivery	Contract
600,000 barrels	Brent System	June 1985	NEP 35

and whereas A and B and C and D and E have concluded and agreed that it would be in their joint best interest to terminate each of the contracts identified above.

Now therefore it is hereby agreed as follows:–

1.   In consideration of the execution of this agreement to terminate each of the contracts identified above and entering into this agreement to pay the cancellation fees as set forth below and subject as hereinafter provided each party hereby expressly releases each other party and their successors, assigns and legal representatives from all liability, claims and demands arising out of the contracts identified above and each of the above identified contracts is hereby terminated with effect from the date hereof.

2.   That in order to calculate the cancellation fee payable by each party hereto a base price of US Dollars X per barrel ('the base price') shall be used.

3.   Each party hereto shall pay to its seller a cancellation fee equal to 600,000 multiplied by the amount, if any, by which such partys purchase price per barrel exceeds the base price provided always that if the purchase price is less than the base price that party shall receive from its seller a cancellation fee equal to 600,000 multiplied by the difference between the two prices.

Each such cancellation fee shall be paid in US Dollars in immediately available funds on or before 3rd June 1985 and shall be discounted based on a 7.8125 per cent per annum interest rate for 42 days.

4.   All payments hereunder shall be made in full without set-off, deduction or counterclaim.

5.   This cancellation agreement shall be construed and interpreted in accordance with the laws of England and shall constitute the entire agreement between the parties.

6.   The signatories hereto hereby warrant and represent that they are authorised and empowered by their company to consent to this agreement.

7.   This agreement shall come into effect when company A has given telex notice to all other parties to this agreement, stating that A has received the telexed agreement of all such parties to the terms of this agreement.

In witness whereof this agreement has been entered into the day and year first written above.

By company A   By company D
By company B   By company E
By company C

Unquote

The following telex should be sent to company A and all other participants:

Quote

We refer to the cancellation agreement dated as of 22nd May 1985 by and among company A and company B and company C and company D and company E (Book-out ref ABC nn). By this telex we, (insert company) confirm our agreement to the terms of the said cancellation agreement and further confirm that we have calculated and agreed with our seller the discounted cancellation fee to be payable pursuant to clause 3 above.

We further agree to pay, or, as appropriate, receive payment of the cancellation fee on or before the agreed date.

Company Name
by

Unquote

Grateful your confirmation soonest.
Regards

CHAPTER 13

THE GROWTH AND STRUCTURE OF THE BRENT MARKET

13.1 Introduction

We have described the historical origins and the staged development of the Brent market, and explained the methods and rules of its operation. We now turn to a study of the size, growth and structure of the market. To borrow a medical analogy, Chapter 11 was concerned with the embryology, and Chapter 12 with the physiology: this chapter will be concerned with the anatomy of the Brent market.

We shall attempt to assess the size of the Brent market in terms of both the number of transactions and the number of participants, and to measure the growth in activity during 1984–5. Since Brent dealings are both wet and paper, it will be useful to estimate the volume of wet dealings and the multiplied effect of paper dealings on total activity. This effect is manifested by the length of the daisy chain.

It is useful to recall at this point that the Brent market involves spot and forward trading of the Brent blend. Transactions therefore vary in their degree of forwardness, and there is a term structure of deals. We shall measure changes in the composition of deals with different maturities, and analyse changes in the term structure in order to cast light on the relative prices of deals for different forward months made on the same day.

Participants in the Brent market are not restricted to primary sellers and end-users of the Brent blend; they include other oil companies and oil traders. The composition of participants and their identity, the distribution of deals among them, and hence the degree of market concentration are interesting aspects of the market which we shall investigate.

The motives of participants go beyond the simple objectives of trading: the need for a seller to dispose of a barrel of oil produced or drawn from its inventories, or the need for a refiner to acquire a barrel for processing. Trading in Brent, as in other commodities, also involves hedging against uncertainty in a particular market, cross-hedging against uncertainty in other markets and speculating about the differences between prices at different dates. The role of hedgers and speculators is therefore worth studying and this chapter briefly considers the issues involved.

13.2 Data

The analysis of the structure of the Brent market undertaken in this chapter is based on the following sources:

— data kindly made available to us, on a confidential basis, by *Petroleum Argus* for the purposes of this research
— interviews with the staff of some forty companies most of whom are directly involved in the Brent market
— information provided by some of the sponsors of the study

Because of the confidential nature of most of these sources, the presentation will concentrate on the main statistical features of the Brent market, and will leave out specific details about particular transactions. Companies will not be named when the analysis is about individual conduct, but they will appear in lists referring to groups of participants in the Brent market. In some instances it is necessary to be specific lest the story lose too much substance. In all such instances, the information is derived from interviews and not from *Petroleum Argus*.

A short description of the *Petroleum Argus* data is in order, because the interpretation of results depends critically on both the strength and the shortcomings of the sources used. *Petroleum Argus* keep detailed records of spot crude trading as a basis for their daily oil market report and as part of their weekly summary of the crude oil market published in the newsletter *Weekly Petroleum Argus*. They take particular care in recording Brent operations, distinguishing carefully between 'dated' and '15-day' contracts. The daily work-sheets record the type of contract, the price and the names of the seller and the buyer (if known). These records begin in late July 1983 and continue to the present.

We have reprocessed the worksheets to provide a series of monthly trading matrices. Each matrix records for a calendar month and for each active company the number of sales and purchases of a Brent contract (identifying in separate columns contracts for the current and for one, two and three months forward). Sales and purchases are then totalled by month and by company, and the net position for each company in the month can be calculated. The matrix as a whole should sum to zero.

Table 13.1 shows the number of transactions recorded by *Petroleum Argus* in 1984–5 in each quarter. The *Petroleum Argus* data identify participants for a large proportion of deals. For 1984 the information on who precisely bought or sold is available for 87.1 per cent of the deals recorded. It is rare to find sources on trade transactions for any commodity that provide such detailed, and almost complete, information about participants.

Table 13.1: Characteristics of *Petroleum Argus* Data

	1984				1985		
	1Q	*2Q*	*3Q*	*4Q*	*1Q*	*2Q*	*3Q*
Number of recorded transactions	450	552	845	796	1130	1025	1264
Total number of recorded sells and buys	900	1104	1690	1592	2260	2050	2528
Unidentified parties to transactions	81	115	210	274	404	259	328
Identified parties to transactions	819	989	1480	1318	1856	1791	2200
Identification rate (%)	91.0	89.6	87.6	82.8	82.1	87.4	87.0

13.3 The Growth of the Brent Market

Brent market trading (Table 13.2) appears to have grown much more rapidly than Brent system production during 1984. The number of recorded transactions increased by 60 per cent in the second half of the year compared with the first half, while Brent system production increased by only 5.8 per cent. In the first quarter of 1985 the number of recorded transactions rose very significantly once again. There was a 70 per cent increase over the 1984 average, and a 150 per cent increase over the corresponding quarter in 1984. This level of activity was sustained in the second and third quarters of 1985.

This expansion of the Brent market, starting in mid-1984, happened at a time when a marked decline in spot oil prices was

Table 13.2: The Brent Market. Transactions per Quarter. 1984 and 1985

	1984				1985		
	1Q	2Q	3Q	4Q	1Q	2Q	3Q
Number of recorded transactions	450	552	845	796	1130	1025	1264
Volume of Brent blend produced per week (mb)	6.6	5.5	6.0	6.8	6.9	5.9	5.9
Volume of Brent blend liftings per week (mb)	6.2	5.1	5.5	6.2	6.2	5.0	5.3

taking place. It is too early to say whether this was pure coincidence, or whether there was a causal relationship in either direction between falling prices and growth in spot/forward transactions.

Although the quarterly data on the total number of recorded deals reveals a growth in the use of the market with significant jumps between the second and third quarters of 1984 and between the fourth quarter of 1984 and the first three quarters of 1985, there are in fact important fluctuations which are revealed if we analyse the data on a monthly basis.

Using published data from *Argus* for the number of deals made *in each month* for the period September 1983 to September 1985, we obtain the series shown in Figure 13.1. It can be seen that the general steep rise in activity conceals sharp decreases occurring on a temporary basis. There was, for example, a sharp fall in December 1984 and January 1985 followed by an exceptionally large rise in February 1985.[1] Clearly there are a number of forces at work in determining the total amount of activity in the Brent market.

13.4   The Size of the Brent Market

The data at our disposal give a clear idea about the growth of the Brent market but they do not indicate exactly its overall size. Information gathered during the course of our interviews leads us to believe that the *Petroleum Argus* data cover approximately two-

[1] Note also the fall in the number of deals in April 1985 and the sequence consisting of a big drop followed by a very large increase in July–August 1985.

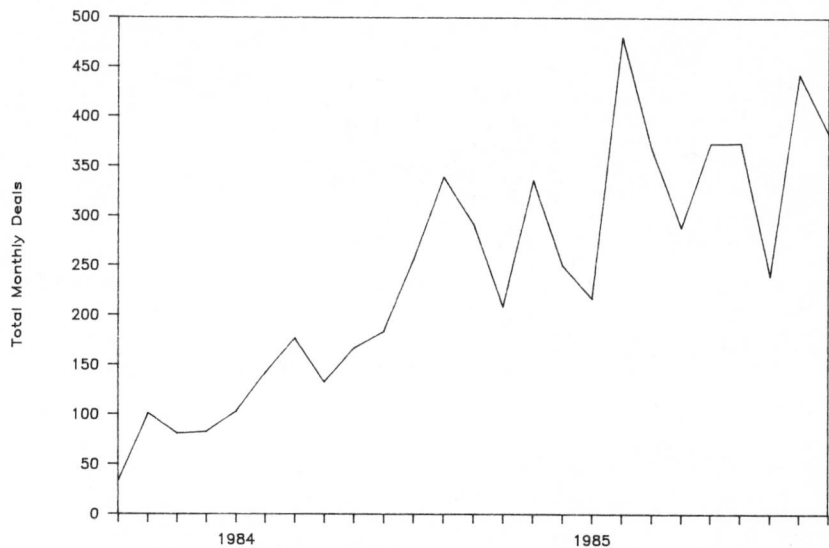

Figure 13.1   Total Monthly Deals for Brent. September 1983–September 1985

thirds of the Brent market. If this estimate is correct, the actual number of transactions would be 675 in 1Q84 rising to 1896 in 3Q85. The total number of transactions in 1984 would be almost 4000. However, it is also important to form an impression of the relative numbers of paper and wet transactions as this provides a better guide to the extent to which the Brent market is being used for trading oil rather than for speculative purposes.

The volume of wet transactions may be estimated, albeit imperfectly, by an indirect method. The *Petroleum Argus* data enable us to identify the participants that tended to remain long and those that remained short during 1984. The overall level of net sales and purchases, properly adjusted and interpreted, can provide an indication of the volume of wet barrel trading.

The method adopted can be described as follows. The first step is to list all primary sellers of Brent blend (List 1, Table 13.3). The second step is to list other companies that appear to be significant net sellers and that are known or thought to have had BNOC contracts or to have acted as agents for BNOC. These companies could have had primary access to Brent blend (List 2, Table 13.4). These two lists include all potential primary sellers of Brent blend identified in our sources. The third step is to work out the net

Table 13.3: Companies with Primary Availabilities of Brent System Crudes (List 1). Availabilities in Thousand Barrels per Day. 1984.

Company	Fields	Availabilities
Amerada Hess	Hutton, NW Hutton, Murchison (Norway)	5.1
Amoco	Hutton, NW Hutton, Murchison (Norway)	14.1
BNOC	All fields, Royalty and Participation plus Assignments	200.2
Britoil	Dunlin, Hutton, Murchison (UK), Thistle	23.4
Conoco	Dunlin, Hutton, Murchison (UK and Norway)	30.8
Deminex	Thistle	15.5
Esso	Brent, Cormorant N and S, Dunlin, Murchison (Norway)	265.4
Gulf	Dunlin, Hutton, Murchison (UK)	28.3
Mobil	Hutton, NW Hutton, Murchison (Norway)	14.5
Saga	Murchison (Norway)	0.5
Santa Fe (Kuwait)	Thistle	6.3
Shell	Brent, Cormorant N and S, Dunlin, Murchison (Norway)	265.4
Statoil	Murchison (Norway)	14.6
Texas Eastern	Hutton, NW Hutton, Murchison (Norway)	3.0
Tricentrol	Thistle	3.7
Ultramar	Thistle	0.5
Total		891.3

Note : Burmah, Charterhouse and Enterprise Oil assign their oil. They own shares in Brent system oilfields but are not primary sellers. Some of the companies were not active on the Brent market in 1984.

Table 13.4: Other Companies with Indirect Primary Access to Brent System (List 2). 1984.

Gotco
Idemitsu, Internorth/P & O Falco
Marc Rich, Mitsubishi
Phibro
Scan, Sohio
Voest Alpine

Table 13.5: The Brent Market. Estimates of Net Sales. 1984 and 1985

	1984				1985		
Number of Sales	*1Q*	*2Q*	*3Q*	*4Q*	*1Q*	*2Q*	*3Q*
1. As Recorded by *Argus*							
(a) By Primary							
Sellers (List 1)	17	6	54	49	41	26	33
(b) By Traders (List 2)	20	22	28	34	17	5	25
Total (Lists 1 and 2)	37	28	82	83	58	31	58
2. As Estimated							
(a) Actual number							
of wet sales =							
1.5 (List 1 + List 2)	55	42	123	124	87	46	87

position of these companies for each quarter. The results are shown in Table 13.5.[2]

Because of imperfect coverage, the actual volume of wet barrels on the spot/forward Brent market is certainly higher than the volume of net sales as recorded by *Petroleum Argus*. We shall assume that net sales are underestimated by the same factor as total transactions. The estimate of actual net sales is then made according to the following formula:

Actual net sales = 1.5 (List 1 + List 2)

The results should be cautiously interpreted as indicating an order of magnitude given the tentative nature of the assumptions made.

13.5 Daisy Chains

The estimates of wet sales on the spot/forward Brent market can be used for two purposes. First, they may be compared with total production and total liftings of Brent blend crudes in order to

[2] The method is to compute the number of deals made by the relevant companies *for* each month and to aggregate these numbers to obtain the net position *in* the relevant quarter.

calculate the proportions in which Brent has been disposed between the spot market and other types of transaction (term sales and internal appropriations by integrated producing companies).

Secondly, the number of wet deals may be compared with the total number of transactions in order to obtain an estimate of the average length of the daisy chain. The results of these various calculations are shown in Table 13.6.

It appears from Table 13.6 that the increase in the total number of transactions in the Brent market in the second half of 1984 was accompanied by a very large increase in the volume of wet deals: the growth of the Brent market in that period stands firmly on an expansion of wet barrel trading. The proportion of total Brent liftings from Sullom Voe that was disposed of in the Brent market, instead of being sold in term contracts or directly appropriated, rose from an estimated average of 33 per cent in the first half of 1984 to an average of 81 per cent in the second half of that year; however, this proportion fell again in the first half of 1985. More oil supplies to the spot/forward market led to increases in the numbers of both paper and wet transactions. In fact the increases in wet deals and total transactions seem to have been of the same proportional magnitude in most of 1984. This is apparent from the estimates of average daisy chain length, which seem to remain fairly constant at around 10 in all but the second quarter of 1984. Significant increases in the length of the daisy chain occur in 1985 when there is a quantum jump from 16 to 32 in the number of daisies in each chain between the first and second quarters of that year.

It is interesting that the average length of chains did not change in 1984 (except in 2Q) despite of expansion in market activity. The average ratio of paper deals to wet cargoes supplied remained stable. One possible explanation is that the 'paper multiplier' of wet transactions is more a function of participants' behaviour than of the volume of oil supplied to the market. With stable behaviour on the part of participants, the total number of paper transactions will tend to vary directly with the volume of wet sales and the length of the chain will remain fairly constant.

Participants' behaviour may vary for two main sets of reasons:

— New market circumstances leading to a change in the range of participants' perceptions about prices in the future, a different assessment of market risks, or different trading patterns and requirements.
— Changes in the identities of participants or, more precisely, in the composition of their group.

Table 13.6: The Brent Market. Wet Deals and Daisy Chains. 1984 and 1985.

	1984				1985		
	1Q	2Q	3Q	4Q	1Q	2Q	3Q
Total number of recorded transactions	450	552	845	796	1130	1025	1264
Estimated number of actual transactions (in the quarter)	675	828	1267	1194	1695	1537	1896
Estimated number of actual transactions (for the quarter)	558	673	1065	1422	1377	1503	1564
Volume of Brent blend liftings from Sullom Voe (mb)	80.8	66.5	71.7	80.8	81.0	65.2	69.2
Estimated number of wet deals	55	42	123	124	87	46	87
Volume of wet deals (mb) (0.5 mb per parcel in 1984 and 0.6 mb in 1985)	27.5	21	61.5	62	52.2	27.6	52.2
Supply of Brent blend to the spot market as percentage of total liftings	34	32	86	77	64	42	75
Average length of daisy chains	10	16	9	11	16	32	21

We shall see in section 13.7 that the number and identity of participants did not change in any discernible manner during the period considered. The inference is that changed market circumstances had a greater impact on participants' behaviour in the first half of 1985 than they did in 1984.

Finally, it is worth noting that these results on wet deals and length of daisy chains are consistent with what is known from a variety of sources about the market. Most interviewees told us that many daisy chains tend to have 15–20 links. (Observers perceive the mode, that is the most frequent occurrence, rather than the average. In the case of daisy chains, the mode is likely to be higher than the average because some wet deals are spot transactions. This brings down the value of the average length of daisy chains.) Market observers are also agreed that the volume of crudes sold on spot markets increased considerably at some point in 1984; that the amount of 'paper' activity in the Brent market also increased in that year; and that the market became hyperactive in 1985.

13.6   The Term Structure of the Brent Market

A second important dimension of trading in the Brent market is the 'term structure' of the deals. A significant feature of the Brent market, which distinguishes it from most other spot crude markets, is that deals can vary from being for the current month to being for three or even four months ahead. Thus, at any one time, transactions of different degrees of forwardness are made; and the aggregate number of transactions made during, say, a week or a quarter, would in fact consist of three or four types of deals, identical in all respects except as regards the maturity date. It follows that at any one time there are effectively several prices for Brent depending on the degree of forwardness of the deals.

The relative shares of contracts with different dates in total transactions and the relative price differentials between these contracts tend to change continually. We have examined these variations in both shares and relative prices by constructing an index of 'the average forwardness' of all deals made in a month. The index is shown in Figure 13.2 for the period September 1983 to September 1985.

The value of the index varies from around 0.5 months in autumn 1983 to a peak of around 1.5 months in July–September 1984. There is apparently a strong lengthening in the degree of forwardness up to the early autumn of 1984; this, however, is

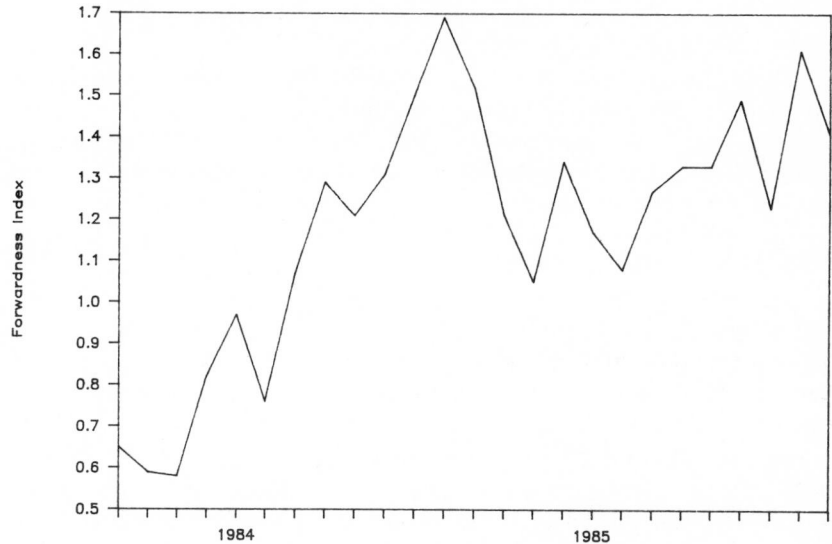

Figure 13.2 Monthly Index of Forwardness of Brent Deals.
September 1983–September 1985

followed in late 1984 and in the first half of 1985 by a shortening in the degree of forwardness, as the value of the index declines to 1.0 month. Since any deal for a month is *on average* for two weeks ahead, we can see that the average forwardness has, in fact, ranged between one and two months.

We then examined the relationship between the forwardness of deals (as expressed by the index) and the following variables:

— the overall level of activity in the Brent market
— the level of physical Brent production

We found that the degree of forwardness appears to be weakly correlated with activity on the Brent market, and that there seems to be no correlation at all with the volume of Brent output. The regression analysis estimates the squared correlation between the index of forwardness and the number of paper deals at 0.37, and the squared correlation with physical production at 0.16. The latter relationship turned out to be negative.

The relationship between the number of deals and the degree of forwardness can also be described by the range in the number of deals of different forward lengths undertaken in each month during the period September 1983–May 1985:

Number of Deals per month	Lowest number	Highest number
Current month deals	22	96
One-month deals	41	263
Two-month deals	14	120

This shows clearly that there is greater variability in the number of forward than current month deals. In other words, the picture is one in which increases in Brent trading are differentially concentrated in forward months.

We then turned to the question of price differentials between transactions with various maturity dates made on a given day. We found that the prices of these transactions are broadly correlated with each other as shown in Table 13.7.

Table 13.7: Correlations Between Brent Prices for Transactions with Different Degrees of Forwardness.

Prices	R^2
Current month and one-month forward contracts	0.975
Current month and two-month forward contracts	0.909
One-month and two-month forward contracts	0.974

We also calculated the differentials between 'forward' and 'current' prices, the latter being defined as the monthly average price for transactions for the current month (the nearest there is to a spot price). The values of the one- and two-month price differentials obtained by subtracting the relevant forward price from the spot price are shown in Figures 13.3 and 13.4 respectively. Commodity traders describe periods in which spot prices are above forward as being in 'backwardation' and periods in which spot prices are below forward as being in 'contango'.

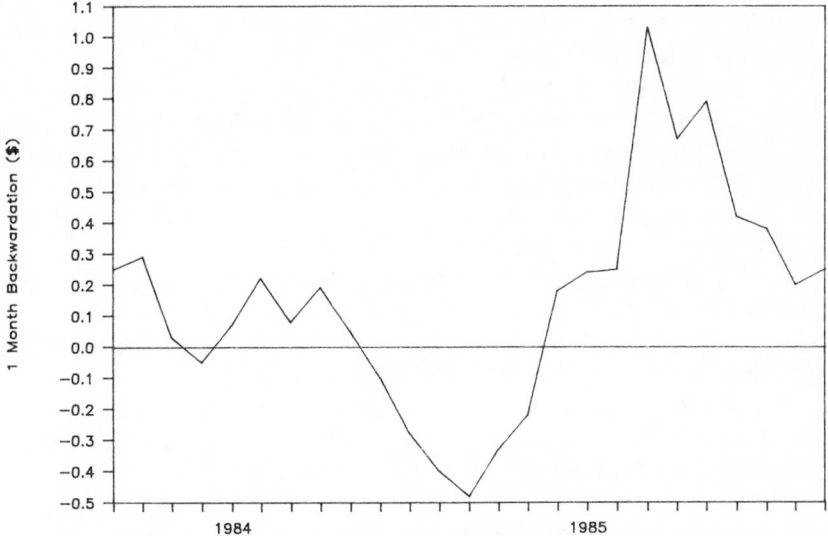

Figure 13.3   Average 1 Month Backwardation (Spot–Forward).
Brent, September 1983–September 1985

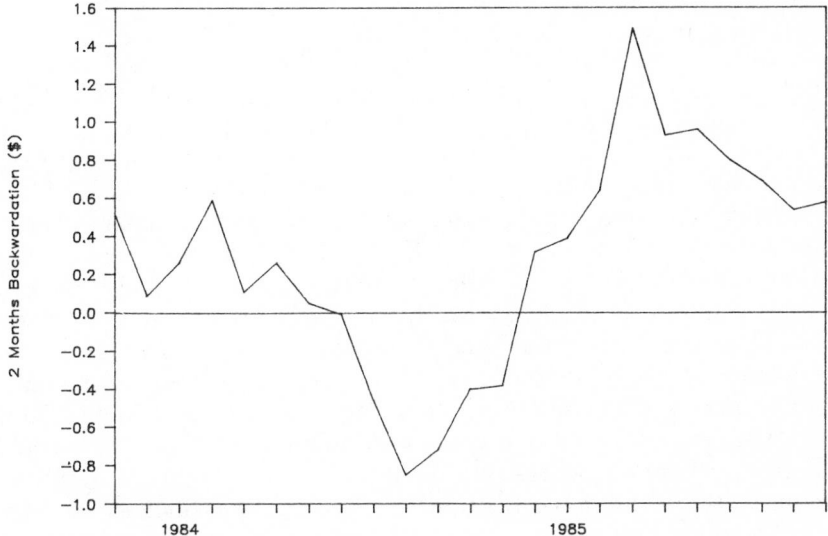

Figure 13.4   Average 2 Month Backwardation (Spot–Forward).
Brent, November 1983–September 1985

Both series show a very similar picture – from September 1983 until May 1984 there was a backwardation on both the one-month and the two-month differentials (spot above forward) with the one-month backwardation averaging around 15 cents and the two-month averaging around 30 cents. There was then a very sharp cross-over of prices producing a contango (spot below forward) peaking around August/September 1984 for one month at 45 cents and two months at around 80 cents. The contango then declined until the market returned into backwardation by the beginning of 1985. The one-month price differential was always around one half the value of the two-month price differential.

This very strong pattern is clearly correlated with the degree of 'forwardness' of trading. The index of forwardness was at its greatest in August/September 1984 and this is also the period when forward prices were most above spot prices.

We found that the index of forwardness is correlated with both the number of deals and the one-month price differential, but the correlation with the price differential is negative. This is a particularly striking result and indicates that if participants of the Brent market decide, for some external reason, to trade further forward, then the price differential (spot less forward) narrows.

13.7   Market Composition

Despite the rapid growth of the market during 1984 and 1985 the number of active participants at any one time appears to have stayed fairly constant. We have identified a total of 125 separate trading entities in the *Petroleum Argus* data. These 125 include in some cases companies belonging to the same corporate group (e.g. Royal Dutch/Shell) which operate as autonomous decision-making centres under their own name (e.g. SUKO, SITCO, Deutsche Shell, etc.). Companies belonging to other corporate groups (e.g. BP or Sun) appear under a single name. Depending on the context of the analysis we shall treat the Shell group of companies either as distinct entities or as a single unit. Thus, assessment of how many autonomous or semi-autonomous agents operate on the Brent market should include the Shell companies under their various names, while an appraisal of the relative weight of large oil companies in the market should aggregate all the Shell subsidiaries into a single unit.

The 125 firms, the total population of the *Petroleum Argus* data set, were not all active throughout 1984–5. The number of

Table 13.8: Brent Market Participants. Buyers and Sellers. 1984 and 1985

	1984				1985		
	1Q	*2Q*	*3Q*	*4Q*	*1Q*	*2Q*	*3Q*
Total number of participants	68	69	75	72	72	70	83
Gross sellers	56	54	54	64	60	63	70
Gross buyers	59	60	71	65	62	66	79
Net sellers	29	31	28	28	35	26	24
Net buyers	30	33	40	30	33	32	43

participants active in any quarter ranged between sixty-eight and eighty-three (see Table 13.8). Of these, forty-two companies were active in every quarter of 1984, and thirty-two of these continued to be active in 1985. Thus, considering the period as a whole, it is possible to divide the 125 identified participants into:

— thirty-two continually active throughout the seven quarters
— an additional ten, who were continually active throughout 1984, but seem to have dropped out in 1985
— eighty-three companies who traded occasionally during the period

Looking at 1984 as a whole it would seem that there were more companies involved in buying than in selling on the Brent market. In 1984 the number of gross sellers (eighty-eight) was smaller than gross buyers (ninety-eight) and similarly net sellers (forty-one) were fewer in number than net buyers (fifty-three). We also found that of the fifty-three net buyers identified in 1984, twenty companies were pure buyers (no sales recorded during the year). Not surprisingly, fifteen of these pure buyers were refiners. The remainder were traders, a feature of the data which may reflect recording failure although it is quite possible that these five traders were just operating on the market as agents of some end-user.

The increases in buyers in 3Q84 (see Table 13.8, lines 3 and 5) and in sellers in 4Q84 are difficult to explain. The majority of new buyers in 3Q84 appear to have been refiners, but the additional sellers in 4Q84 were already in the market as buyers. One possible explanation is that more refiners entered the market in the second half of 1984 initially as simple buyers, and then subsequently became more sophisticated in their use of the market and engaged in selling, as well as buying, for hedging or speculative purposes.

Participants in the Brent market may be classified in the following groups:

(a) Integrated companies with primary availabilities of North Sea crudes (Producers/End-users)
(b) Non-integrated companies, including BNOC, with primary availabilities of North Sea crudes (Producers/non-End-users)
(c) Refiners in North West Europe and the USA with no primary availabilities of North Sea crudes (non-Producers/End-users)
(d) Companies with no primary availabilities of North Sea crudes and no refineries in North West Europe or the USA (non-Producers/non-End-users).

The forty-two continually active participants of 1984 included:

— seventeen companies from group (a) (Producers/Users)
— two companies from group (b) (Producers/non-Users)
— five from group (c) (non-Producers/Users)
— eighteen from group (d) (non-Producers/non-Users)

It is interesting to note that the companies in Groups (a), (b) and (c) include almost all the significant primary sellers and end-users of North Sea crudes identified in Chapter 3; the exceptions are: ICI, Getty, Lasmo, Veba, the US SPR and, until 1985, Statoil and Svenska Petroleum.

Comparing the number of participants in each group with its share of total activity suggests that the average level of activity in each group is very similar. This feature, however, is deceptive. The proportion of participants continually active in each group is much higher for integrated and non-integrated companies with primary access to North Sea crudes (50 per cent) than for traders (33 per cent). Clearly, the small group of traders who are continually active on the Brent market commands a much higher share of total activity than the primary sellers (Groups (a) and (b)).

A number of general conclusions can be drawn from this analysis:

— Both the size and the composition of the market have remained fairly constant throughout 1984.
— Trading companies account for the lion's share of the Brent market (50 per cent of identifiable business).

Table 13.9: Brent Market. Number of Participants by Group. Percentage Share of Activity. 1984

	Number of Participants		% Share of Activity
	Total	Continually Active	
(a) Integrated	37	17	35.0
(b) Non-integrated	4	2	5.1
(c) Refiners	13	5	8.7
(d) Others	54	18	51.2
Total	108	42	100.0

— The group of trading companies is the most volatile in composition since only one-third of this group is continually active compared with a typical 50 per cent of continually active participants in other groups.
— Integrated companies (group (a)), which need not sell or buy large quantities except for tax reasons or crude portfolios, account for a large share of activity in the Brent market (35 per cent), while non-integrated sellers and refiners, which need to dispose of their supplies to third parties or obtain from them their requirements, account for a much smaller share (14 per cent).

13.8 Market Structure

The striking feature of the Brent market is the concentration of activity in the hands of a relatively small number of participants. The top ten companies account for 53 per cent of all recorded deals, the top twenty for 75 per cent and the top thirty for 87 per cent. This means that the scale of activity of some seventy-eight participants in 1984 (108 minus the top thirty) was extremely limited.

One method of quantifying the degree of concentration is to calculate an index known as the Gini coefficient. This is a measure of relative inequality. A Gini coefficient takes values between zero and one. The closer the value to zero, the greater the degree of equality between participants' shares of activity and the smaller the degree of concentration. Accordingly the higher the value of the Gini coefficient the greater the degree of concentration.

Table 13.10: Concentration of the Brent Market. Gini Coefficients, 1984 and 1985

	1984				1985		
	1Q	*2Q*	*3Q*	*4Q*	*1Q*	*2Q*	*3Q*
Number of identified transactions	819	989	1480	1318	1856	1791	2200
Number of participants active in the quarter	68	69	75	72	72	70	83
Average number of transactions by participant	12	14	20	18	26	26	26
Gini coefficient	0.52	0.58	0.61	0.60	0.60	0.61	0.62

Gini coefficents were calculated for each quarter. The results, which are presented in Table 13.10, confirm that the distribution of Brent trading activity is highly skewed in favour of the most active participants. The value of these coefficients tended to rise between 1Q84 and 3Q84 and then stabilized. This rise coincided with a large increase in the number of transactions and a small rise in the number of participants active in successive quarters. The conclusion is clear. As the market became more active the degree of concentration increased. The increased volume of trading activity was not shared more or less equally by all participants but took place among the members of the small group at the top of the list. The degree of concentration (or the inequality in the distribution of transactions between participants) is illustrated fairly vividly by Figure 13.5.

The ten most active participants in the Brent market in 1984 included five oil companies and five traders. Two of these five oil companies have no primary availabilities of Brent system crude. More revealing is the fact that, out of the five most active participants in 1984, four were traders.

Interviewees provided us with their own lists of main participants in the Brent market. The quasi-unanimous view is that among oil

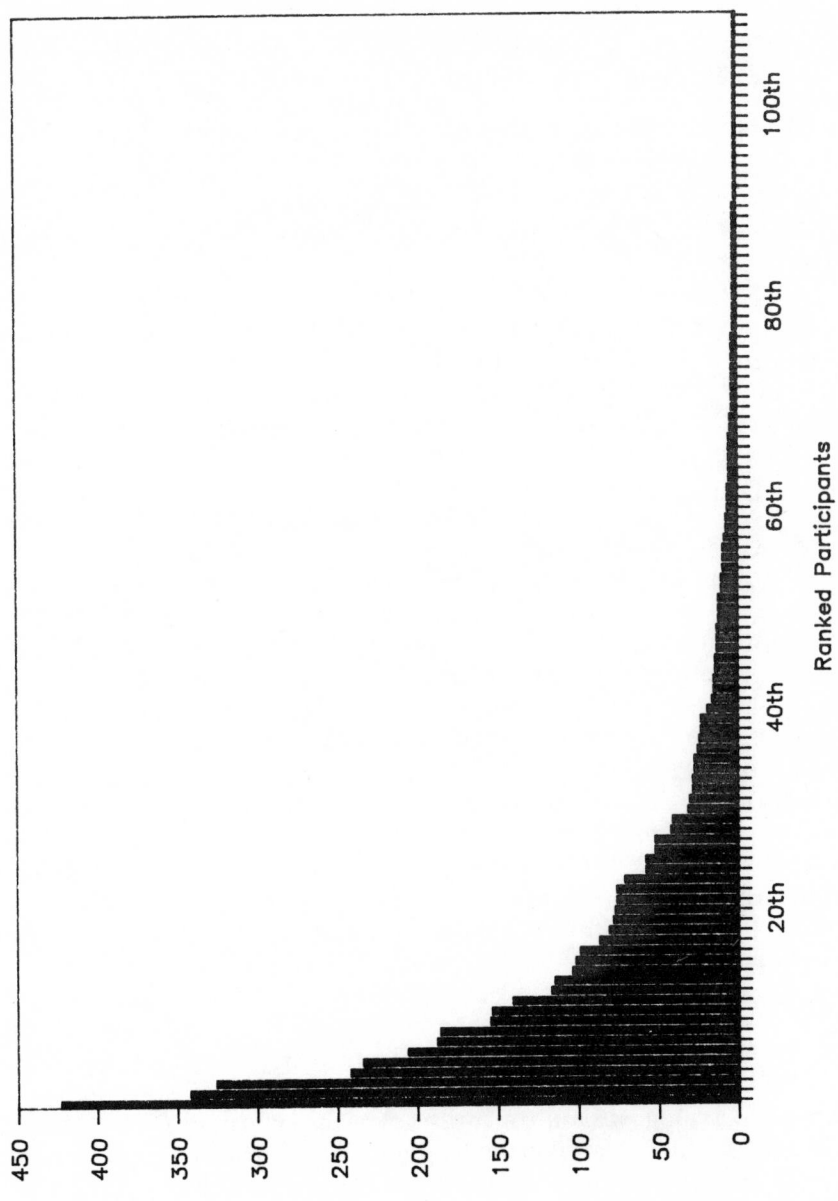

Figure 13.5   Deals per Participant on the Brent Market. 1984

companies BP had the highest profile and carried considerable influence as a result of this general perception of its role.

Interviewees also indicated that Sun was a very large buyer. This is confirmed by US trade statistics. Other oil companies said to be active on both sides of the market were Gulf and Shell. Traders frequently named were Phibro, Transworld Oil, Voest Alpine, Internorth, P & O Falco and Gatoil.

There is no evidence that BNOC was among the very top players in terms of total deals, but was clearly a top net seller, particularly in the second half of 1984.

The *Petroleum Argus* data also show that the ten top net sellers included five oil companies, three of which have Brent blend availabilities, and five traders. The ten top net buyers included four oil companies with refineries (potential end-users) and six traders. The presence of producers among the net sellers and of refiners among the net buyers is to be expected. However, the strong presence of traders at the top of both lists is surprising. The size and regularity of net sales and purchases in all these cases tends to argue against the simplest explanation: under-reporting by *Petroleum Argus* of a kind that systematically biases the results.

It seems that traders who were net sellers had term availabilities from BNOC or from producers of Brent system crudes, or were acting as agents on behalf of these primary sellers. Voest Alpine and some of the Japanese traders (e.g. Mitsubishi) definitely fall into this category. Traders who were net buyers may have had processing arrangements with European refiners, or owned shares in refineries (e.g. Gatoil, Scan) or may have been purchasing on behalf of end-users.

An important result of this analysis is that neither primary availability of Brent (nor for that matter of other North Sea crudes) nor end-use requirements provide a good criterion for identifying active participants in the Brent market. This can be illustrated by the following:

— Out of the ten top European refiners of North Sea crudes, only three were among the ten top Brent market participants.
— Out of the ten top US importers of North Sea crudes only two were among the ten top Brent market participants.
— Out of the ten top primary sellers of North Sea crudes, only three were among the ten top Brent market participants.
— Out of the ten top Brent blend producers, only three were among the ten top Brent market participants, and no Brent blend producer was to be found among the top five Brent market participants!

13.9 Hedging and Speculating in the Brent Market

Having described the structure of the Brent market both in terms of the type of transactions and the type of participants, it is now useful to analyse the motives for trading, in particular hedging and speculating.

We shall define these terms in a very specific way and must therefore guard the reader against misinterpretations of our analysis that may arise from the broad meaning of these words in common usage. We define a hedger as an agent who has a commitment to buy (or sell) the physical commodity on a regular basis at a market price. In the Brent market, hedgers in our sense of the term include only primary sellers and end-users of Brent. Given the uncertainty attached to future prices the hedger faces an uncertain revenue (if selling) or cost (if buying); and his attempts to reduce this future uncertainty by entering into forward deals are termed hedging. The essential feature of a hedge is that it represents a move from a higher level of uncertainty to a lower one.

In contrast, the speculator has no prior commitment to buy (or sell) the physical commodity. He only buys (or sells) because he expects to be able to sell (or buy) later at a profit. In a sense, the speculator moves from a position of certainty (no requirement to trade, hence no risk and no profit) to a position of uncertainty when he enters into a deal since he cannot be sure about the future price.

Under these definitions producers of crude could be pure hedgers (if they sold some of their crude forward but entered into no other deals), pure speculators (if they only bought and sold paper barrels), or both. The essential feature of the market is that only the sale (or purchase) of wet barrels can be for hedging purposes (although it might be purely to exchange for another preferred crude).

As we explain in Annex 3, hedging wet barrels creates a 'kitty' with which speculators can be rewarded, and it is the size of this kitty that governs the long-run size of the paper market. The larger the number of wet barrels that are hedged and the greater the margin that hedgers will pay for certainty, the greater the number of paper deals that can be supported. In Brent market terms, the length of the daisy chain is in fact fixed by the margins paid by hedgers and the margin required to reward speculators, and is invariant in the long run with respect to the amount of physical crude put on the market.

The other aspect of a market that determines activity is the range of views held by agents. If all agents have the same skills and needs then there is no gain from trade unless they have sufficiently different beliefs to make them mutually willing to trade at a price. This aspect is of central importance in understanding the operation of a speculative market such as Brent. Given agents who are willing to buy *or* sell, then one will buy and the other will sell at a given price only if both think they will make a profit (one believes the price will fall relatively and the other that it will rise). Given that both need a certain margin to do a risky and costly deal we have the fundamental result that the *wider* the dispersion of views about expected prices the more deals will be done, as more agents can find partners whose views are sufficiently different from theirs.

These remarks allow us to interpret some of the facts that we established earlier for the market:

— We expect the length of the daisy chain to remain fairly steady over the medium term unless the market develops better ways of transmitting information that would, in effect, reduce the premium paid by hedgers for certainty. Of course the length of the daisy chain could vary in the *short run* as a result of changes in wet crude traded or unusual divergence of beliefs; but these variations would be expected to stabilize, as the costs of hedging rose or the profits from speculating fell and as views diverged less.

— The number of deals in the market will also tend to be stable over the *medium term* if the amount of wet crude traded is fairly constant; and they will tend to increase in proportion to wet supplies. However the adjustment to changes in the volume of wet supplies to the market may be asymmetrical. Participants may be slow to reduce their activity when this volume decreases because they may confuse the signal of over-capacity in trading – which is a drop in profit – with a run of bad luck.

— In the short run the number of deals will be sensitive to changes in the range of expectations and not to the general level of expectations. If everybody expects a price fall then nobody will be willing to buy now at a higher price. Thus sudden swings in the level of paper activity on the Brent market (such as in August 1984 or February 1985) are related to unusually wide *divergences* of opinion as to price movements. Lows in activity correspondingly relate to periods of unusual *unanimity* of views.

— In the short run, while players are still learning to play this game, there are opportunities for more skilful (or better informed) companies to speculate against their less skilful competitors and systematically to make a profit over time. Eventually the less skilful market players will improve and will not be prepared to trade in the same circumstances, or else they will drop out of the market. Hence there may be an extra source of revenue for some companies, and a larger number of deals, than could be sustained in the long run. Once the adjustment has occurred, pure speculation will no longer pay *on average* over time for any individual company and the number of speculative deals will drop until they are supported solely by the margin that can be extracted from the hedgers.

A final way of testing how the market is being used is to look at the actual forecasting ability of the forward price. If the market were being used solely for hedging then the price for a deal (say) one month ahead should represent the best gross available of what the future will bring as it should encapsulate *all* the information currently available about the future. In particular the (say) one-month price should be a better prediction of the actual outcome of prices in a month's time than any other price currently available, and no other variable should be able to improve its performance. This is known to economists as the 'efficient markets hypothesis' and is described in detail in Annex 4.

The results of our statistical analysis show the opposite effect. Although this month's *forward* price does predict next month's spot price fairly well, its performance is significantly improved by adding this month's *spot* price. Furthermore this month's spot price alone predicts next month's spot price well, and this month's forward price adds *nothing* significant to the relationship. This strong result indicates that the trading forward in Brent is not primarily genuine hedging in which the actual price in a month's time is the crucial magnitude. Rather the prices in a few days' time are the ones that will determine whether or not the paper transactions will be profitable. Given the observed lengths of the daisy chains it is clear that the market is dominated by deals with a speculative element and such agents cannot afford to hold open positions for lengthy periods of time. It is in their interest to 'pass the parcel' quickly: hence they are dominated by expectations of price changes over the short run.

It should be noted that with companies both hedging (selling or buying wet barrels) and speculating (selling or buying paper barrels) it is not possible to say, given the nature of Brent deals, which particular transactions will turn out to be wet (and hence hedged) and which to be dry (and hence speculative). These distinctions are used as an analytical device from outside the market rather than as a description of how agents perceive the situation.

13.10   Conclusions and Implications

The evidence on the term structure of the Brent market shows that, although activity has generally been increasing on trend, there have been substantial monthly fluctuations in the number of deals.

There were even larger fluctuations in the degree of forwardness of the average deal. Thus, while activity on the market varied, the typical length of a deal varied even more.

At the same time, the price differential between deals for current and forward months also varied. We had a backwardation in late 1983 and in the first half of 1984 which changed dramatically to a contango in August/September 1984, and then reverted to a backwardation in early 1985.

Both these features of Brent market developments suggest that exits from, and entries to, the market were more heavily concentrated at forward dates. Furthermore, exits of sellers (entries of buyers) at the forward end were not associated with spot prices rising relative to forward prices. On the contrary, we found that a relative increase in forward buying was correlated with forward prices rising relative to current prices.

Since changes in the relative levels of activity at the spot and the forward ends of the market do not appear to have been related to the normal economic response to changes in price differentials, we can only infer that another factor was at work. This other factor caused the change in activity levels, which in turn produced the change in price differentials.

This other factor can be identified as 'very short-term trading'. Participants in the Brent market hold their positions open for very short periods. Evidence for this behaviour is obtained by comparing the average length of the daisy chain (10–15 links) with the average degree of forwardness (1.0–1.5 months), a comparison which indicates that the average position remains open for only a few days. Traders seeking to buy (or sell) with a view to sell (or buy) within a few days would be much more interested in the possibilities of price movements over this short period of days than in the price differential (spot/forward) over the medium term. For these traders the forward price as such is less important than the perceived short-run movement in that price. Hence agents do not increase the level of their activity when the forward price falls below the spot price, but rather when they entertain the thought that the forward price will soon move again.

This interpretation also suggests why much of the variation in the level of activity was concentrated at the forward end of the market. When the trader's aim is to speculate against price changes then buying a current month deal is more risky (all other things being equal) than buying further forward. The more time there is available before 'the parcel is passed on' or an actual delivery of unwanted oil is to be taken, the greater the opportunity of finding someone else with whom to do a profit-making deal. In normal circumstances, we would expect speculators to buy (or sell) the longer-dated deals to a greater extent than primary sellers and end-users.

A particularly dangerous situation for the market would arise if the degree of forwardness of deals were to decrease and the daisy chain length to remain fixed. A long daisy chain and lack of sufficient forwardness mean that the time available to agree the chain is reduced, and that the risk of an unwilling agent being required to find physical oil to buy or sell (or to default) will increase. Low forwardness and a high paper multiplier can cause problems for the market. This seems to have happened in the first half of 1985, when the average length of the daisy chain rose from 10–11 to 15–16 and the average degree of forwardness fell from 1.5 months to 1.0. The consequences were predictable: the number of defaults and litigations increased causing unrest in the Brent market. The most responsible participants have worries about this situation which have led them to suggest, very discreetly indeed, the need for some self-imposed but formal regulation.

Table 13.11: Brent Market. Participants Classified in Groups. 1984 and 1985.

Group (a) : *Integrated companies with primary availabilities of North Sea crudes*

Agip, Amerada Hess, Amoco
BP
Caltex, Chevron, Conoco
Dow
Elf, Esso
Gulf
Kerr McGee, Kuwait Petroleum Corporation
La Gloria/Texas Eastern
Marathon, Mobil, Murphy
Norsk Hydro
Occidental
[Petrofina, Petrofina/US], Phillips
[Shell, Deutsche Shell, Norske Shell, Svenska Shell, SITCO, SUKO], Sohio, Sun
[Texaco, Deutsche Texaco], Total
Union, [URBK, Rheinoil]
Wintershall

Group (b) : *Non-integrated companies with primary availabilities of North Sea crudes*

Britoil, BNOC
Tricentrol
Ultramar

Group (c) : *North West European and US refiners with no primary availabilities of North Sea crudes*

Arco, Ashland
Champlin, Charter, Coastal, Crown
Koch
Mapco
Neste, Norol
Petrogal
Saarbergewerke
Tesoro

Group (d) : *Companies with no primary availabilities and no refineries in North West Europe or the USA*

Albaco, Apex, Astra, Astroline, Avant
Bomar, Bonaire
Carey, Cariogi, Citizens Energy, C Itoh, Crysen
Gatoil, Gotco
Horizon
Idemitsu, Internorth/P & O Falco
Kaiser, Kanematsu
Mabanaft, Marc Rich, Marimpex, Marubeni, Menex
Merx Handel, Mitsubishi, Morgan Stanley
Nichimen, Nissho Iwai, Northville, Nova
Oilmen
Pecten, Pegasus, Petrogulf, Petroship, Phibro, Phoenecian, Pilot
Resco
Scan, Sigmoil, Sumitomo

Table 13.11: Continued

Tampimex, Toya Menka, Tradax, Tradinaft, Tricon, TWO
UPG
Vitoil, Voest Alpine
Willco, Willer

Note : The following new names were identified in 1985

Bairsterns, BB Naft
Cedar
Delphi
Goldman Sachs
J Aran
Metallgesellschaft, Mitsui
Natomas
Petronor, Propetrol
Skinner, SNR, Statoil, Svenska OK, Svenska Petroleum
Vanol
+ 3 others

Table 13.12: Top Ten Primary Sellers of North Sea and Brent System Crude. Top Ten End-users of North Sea Crude: European Refiners and US Importers. 1984

Rank	Primary Sellers		End-users	
	Total North Sea	Brent System only	European Refiners	US Importers
1.	BNOC	Shell	Esso	Sun
2.	BP	Esso	Shell	Sohio
3.	Shell	BNOC	BP	SPR
4.	Esso	Conoco	Texaco	Gulf
5.	Statoil	Britoil	Mobil	Amoco
6.	Mobil	Deminex*	CFP	Mobil
7.	Phillips	Amoco	Conoco	Kerr McGee
8.	Conoco	Mobil	Elf	Crown
9.	Petrofina	KPC	Amoco	Diamond Shamrock
10.	Britoil	Gulf	Gulf	Murphy

* URBK, Veba and Wintershall

PART IV

MARKETS AND PRICES

CHAPTER 14

THE PRICES OF NORTH SEA CRUDES

14.1 Introduction

The ultimate purpose of a study of markets is to shed light on the process of price formation. This chapter addresses itself to a set of basic questions on spot price movements, changes in term prices, price relationships between North Sea and major world crudes, and movements in crude oil price differentials.

The objective is to discern regularities in behaviour whenever they obtain. These regularities tend to be concealed by considerable 'noise' in the price data, the inevitable effect of: (a) poor information and statistical errors, and (b) the interaction of a large number of ephemeral factors and events, random shocks and minor economic forces or determinants.

14.2 The Spot and Term Prices of Brent

The movement of the Brent spot price over the period January 1980 to February 1985 is depicted in Figure 14.1. The main feature of this price movement is a steadily declining trend beginning in 1981 associated with significant fluctuations. Oil prices have been gently falling all round the world during recent years, and it is not surprising to observe that the prices of North Sea crudes, represented by Brent, have followed the same tendency.

It would be interesting, however, to measure the rate of decline statistically. An estimate of this rate for Brent obtained by regression analysis (see Annex 5) indicates that the decline was equivalent to an average fall of 18 cents per month over the period considered.

The movement of the BNOC term price for Brent is shown in Figure 14.2. Term prices did not fluctuate continually but were changed by discrete amounts at irregular intervals. In 1980, 1981 and the first half of 1982, term prices were varied more frequently than in the second sub-period (second half of 1982 to mid-1985). In the first sub-period the term price was raised on certain occasions

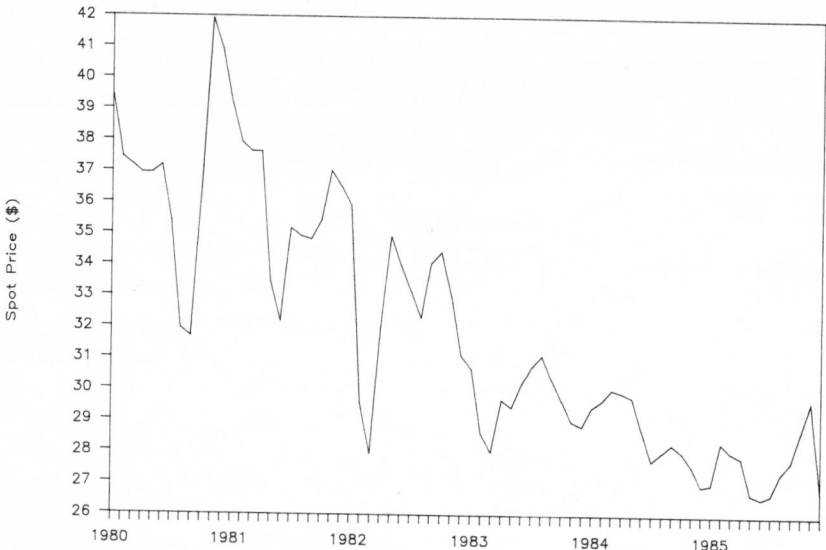

Figure 14.1   Brent Spot Price. January 1980–December 1985

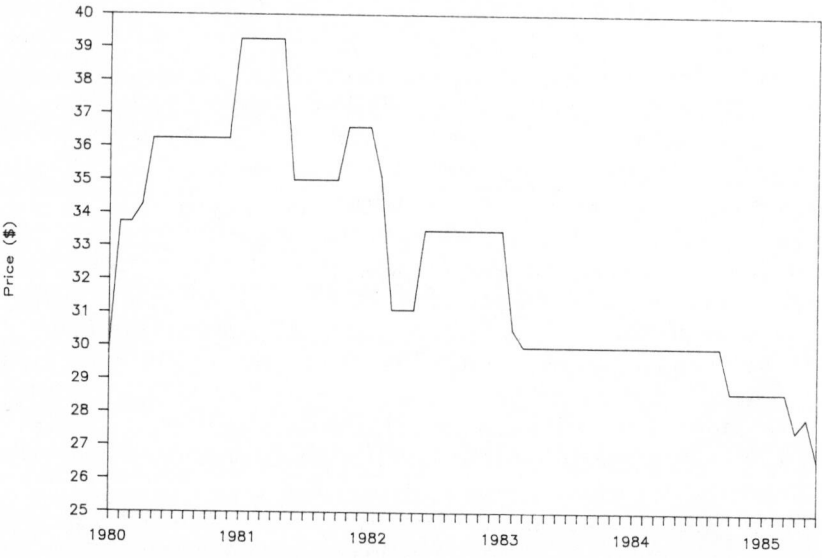

Figure 14.2   Brent Official Price. January 1980–June 1985

and reduced on others, but in the second sub-period the term price was always changed in a downward direction. A remarkable feature of these term price developments was the fairly long episode, stretching over eighteen months between March 1983 and October 1984, during which the Brent term price was kept fixed at $30 per barrel.

The general picture depicted by Figure 14.2 is also one of decline of the term price over the period considered. Applying the same regression technique, we found that the rate of decline of the Brent term price was equivalent to an average drop of 15 cents per month between January 1980 and March 1985.

Thus, the average rate of decline of the Brent term price was smaller than the decline of the Brent spot price. This feature is consistent with a fairly universal relationship between spot and term prices observed in all markets. In periods of boom the spot price of oil tends to move ahead and to stay above the corresponding term or 'official' price; in periods of stagnation or decline the spot price moves, and tends to stay, below the official price. This is partly the result of differences in the mode of price determination: term or official prices are changed at discrete intervals and are bound to lag behind spot price movements in both the upswing and the downswing. Spot and term prices do not equalize because markets are always in disequilibrium, and the dynamic of excess supply or excess demand elicits an immediate spot price response which may or may not be followed by a delayed adjustment of term or official prices. However, the spot price continues to move even when the adjustment is made, and a spot/term differential re-emerges in either direction depending on market conditions.

The comparison of rates of decline (18 cents per month for the spot and 15 for the term price) does not provide an immediate idea of the differential between Brent spot and term prices. Figure 14.3 (which is indeed identical to Figure 8.1 but reproduced again for the reader's convenience) plots the value of this spot/term difference over the whole period January 1980–June 1985. It appears from a cursory glance at the figure that the difference between Brent spot and term prices tended to narrow after mid-1982 compared with the earlier period. Furthermore, spot prices tended to be below term for most of the sub-period mid-1982 to end 1984; the only exceptions to this general tendency being September/October 1982 and June–September 1983.

Regression analysis enabled us to estimate the average differential between the Brent spot and term prices in the period July 1982–February 1985. The result, which is statistically significant,

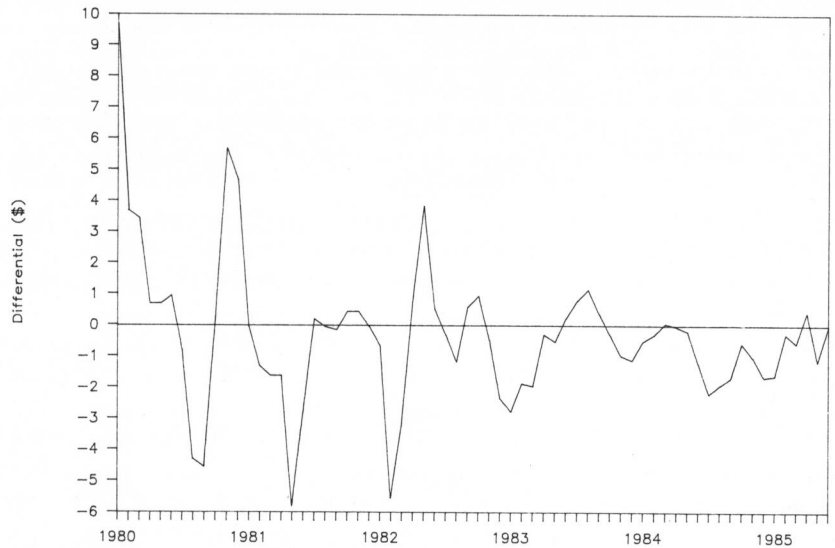

Figure 14.3   Crude Price Differential. Brent Spot/Term, January 1980–June 1985

indicates that the Brent term price was on average 63 cents above its spot level.

This result suggests that BNOC made a maximum (average) trading loss of 63 cents per barrel of participation crude subsequently sold on the spot market by the Corporation. The actual (average) loss was certainly smaller because BNOC did not sell its oil exclusively on the spot market for most of the sub-periods considered. Furthermore, BNOC had the reputation of being a very good trader and may have outperformed the market, realizing a few cents more per barrel than the average spot price.

Our result for the value of the differential between term and spot Brent enables us to calculate the gain (per barrel) made by oil companies on participation crude delivered to BNOC. Between July 1982 and February 1985, these companies made on average 63 cents per barrel before tax and a minimum of 25 cents after tax.

Similarly the net (average) gain from spinning oil for an integrated company is estimated at 24 cents per barrel under the assumptions:

— The company was spinning throughout the sub-period irrespective of the actual sign of the differential.

— The applicable tax rates are 50 per cent for Corporation Tax and 75 per cent for PRT.

Considering the significant magnitude of these per barrel gains, we remain puzzled as to why the oil companies used to express such dissatisfaction with the existence of BNOC and the participation arrangements.

14.3   The Variability of Brent Spot Prices

The Brent spot price not only declined but was also subject to marked fluctuations. A notable feature of Figure 14.1, however, is that these fluctuations became progressively smaller over the period from 1980 to early 1985. An impression of this reduced price variability can be formed with greater clarity by dividing the data into a number of shorter sub-periods. The sub-periods are determined by visual examination of the graph and relate to identifiable price cycles.

A simple indication of volatility is obtained by estimating for each sub-period the error variance of the relevant spot price series (relative to the period mean) by regression on a constant. The results are shown in Table 14.1. They reveal a very distinct pattern of progressively reduced volatility as we move from late 1980 to mid-1984. The standard error of estimate falls from $3.50 around a mean Brent spot price of $36.90 in the first sub-period (September 1980–June 1981) to $0.36 around a mean price of $29.70 in the fifth sub-period (January–June 1984). Variability, however, increases in the second half of 1984. The standard error of estimate rises to $0.46 around a lower mean Brent spot price of $27.85. The variability measure adopted here is the standard error normalized by the mean (the ratio of these two numbers). Since the standard error decreases from one sub-period to the next at a much faster rate than the mean price, it follows that average variability declines. This is shown in the table by an index which takes the first sub-period as a base with an index value of 100.

Before drawing any conclusions from these results based on monthly spot price data, it is useful to measure the volatility of daily and weekly spot prices. The period of interest is 1983–4 when medium-term price volatility seems to have been significantly reduced since this period witnessed the emergence and rapid development of an active forward market.

Table 14.1: Variability of the Brent Monthly Spot Price by Sub-periods. September 1980–December 1984.

Sub-period	Mean Price $	Standard Error of Estimate $	SEE/ Mean	Index
1. Sep 80–Jun 81	36.90	3.50	0.095	100
2. Jun 81–Mar 82	33.96	3.06	0.090	95
3. Mar 82–Mar 83	31.88	2.44	0.077	81
4. Mar 83–Dec 83	29.73	0.94	0.032	33
5. Jan 84–Jun 84	29.70	0.36	0.012	13
6. Jul 84–Dec 84	27.85	0.46	0.017	17

These estimates of short-term variability were carried out on two sets of Brent spot price data:

— Weekly data for the period January 1983–December 1984. The data are end-week prices for 104 weeks taken from *Petroleum Argus*.
— Daily data for the period January 1984–February 1985 taken from *Platt's Crude Oil Marketwire*. The number of observations is 290.

The estimates of variability using weekly data are made for three sub-periods (end March 1983–mid-December 1983, 3rd week December 1983–3rd week June 1984, end July 1984–end December 1984). The results are shown in Table 14.2. They show a decline in variability between the first and the second sub-period (roughly speaking between the last nine months of 1983 and the first half of 1984), and, as for monthly price data, an increase in the second half of 1984.

Table 14.2: Variability of the Brent Weekly Spot Price by Sub-periods. March 1983–December 1984.

Sub-period	Mean Price $	Standard Error of Estimate $	SEE/ Mean	Index
1. End Mar 83–mid-Dec 84	29.90	0.80	0.0267	100
2. 3rd week Dec 83–3rd week Jun 84	29.70	0.45	0.0152	57
3. End Jul 84–end Dec 84	27.66	0.70	0.0253	95

Estimates of variability using daily data only cover 1984 and the first two months of 1985. Three sub-periods are identified with the purpose of removing trend factors (1st January–20th June 1984; 5th July–13th November 1984; 13th November 1984–28th February 1985). The results (Table 14.3) confirm the pattern shown by both monthly and weekly data. They indicate that short-term price variability increased in the second half of 1984 compared with the first half of that year, and that variability continued to increase at the beginning of 1985.

Table 14.3: Variability of the Brent Daily Spot Price by Sub-periods. January 1984–February 1985.

Sub-period	Mean Price $	Standard Error of Estimate $	SEE/ Mean	Index
1. 1st Jan–20th Jun 1984	29.78	0.33	0.0111	100
2. 5th Jul–13th Nov 1984	28.08	0.59	0.0210	190
3. 13th Nov 1984–28th Feb 1985	27.37	0.67	0.0245	221

These results suggest that the beginnings of the Brent market (1980–82) were particularly unsettled, partly because the world petroleum market, at the time, was going through a very difficult process of price adjustment in the aftermath of both the Iranian Revolution and the outbreak of the Iran–Iraq war, and partly because the Brent market itself was still small in size and thin in activity.

The period from March 1983 to mid-1984 appears with hindsight to have been a short 'golden age' of stability. The term price of Brent remained fixed; the market expanded gradually, established its informal rules, and tended to settle. It is not surprising to find that the variability of the Brent spot price declined significantly throughout these particular fifteen months.

In mid-1984, as is now becoming increasingly evident from the results of this study, the Brent market was unsettled. Tax spinning increased, activity in Brent paper deals increased and the spot price declined. We now find that these phenomena were accompanied by an increase in the variability of the spot price.

14.4 Oil Price Differentials: North Sea Crudes

The spot prices of North Sea crudes move in close harmony with each other, and the general tendencies discerned for Brent over the period from 1980 to early 1985 apply to Norwegian and other UKCS crudes. The price movements, however, are not identical: both the rate and the direction of change in the price of North Sea crudes may diverge from time to time. In other words, the price differential between any pair of crudes tends to vary in magnitude, and occasionally in sign, over any relevant period of time.

We shall consider here three North Sea crudes: Brent, Ekofisk and Flotta. Ekofisk is conventionally regarded as the marker for Norwegian crudes; and Flotta is chosen because it is dissimilar to Brent.

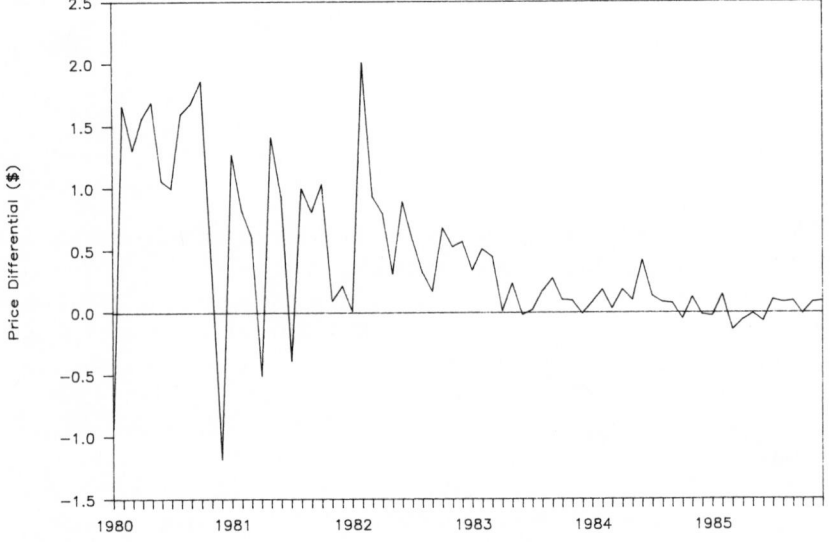

Figure 14.4   Ekofisk/Brent Differential. January 1980–December 1985

The broad closeness in price movements between Brent and Ekofisk is apparent from a cursory look at Figure 14.4. For example, there is almost perfect coincidence in the occurrence of peaks and troughs in the spot prices of Brent and Ekofisk. The coincidence is markedly less perfect in the case of Flotta (see Table 14.4).

A more sophisticated way of assessing the degree of closeness in price movements is to use a regression model. Brent spot prices are regressed on each of the other spot price series (over the maximum data period common to each pair) without a constant. The

Table 14.4: UK and Norway. Peaks and Troughs in Spot Price Movements of Selected North Sea Crudes. January 1980–February 1985.

	1980	1981	1982	1983	1984–5
1. Peaks					
Brent	Nov	Nov	May, Oct	Aug	Mar, Sept
Ekofisk	Nov	Nov	May, Oct	Aug	Apr, Sept
Flotta	Nov	Nov	May, Sept	July	Mar
2. Troughs					
Brent	Sept	June	Mar	Mar, Dec	July, Dec
Ekofisk	Sept	June	Mar	Mar, Dec	July, Dec
Flotta	Aug	June	Mar	Mar, Dec	Jan 85

regression coefficients of the Brent/Ekofisk and Brent/Flotta price relationships are both highly significant in statistical terms. The correlation as measured by R^2 is slightly higher for Brent/Ekofisk (0.973) than for Brent/Flotta (0.957) (see Annex 5).

This broad closeness in the movements of spot prices (Brent, Ekofisk and, to a lesser extent, Flotta) is associated with changes in the behaviour of the relevant price differentials. The interesting question is whether the differential between prices of any pair of North Sea crudes has fluctuated randomly or has changed in a systematic manner over the period considered.

A regression model is used to test whether the Brent/Ekofisk and the Brent/Flotta differentials display regular behaviour. The regression of the Brent/Ekofisk differential on a constant and a trend variable suggests that the price differential between these two crudes tended to narrow continually over the period January 1980–January 1985. The reduction in the absolute value of the price differential is equivalent to an average change of 2 cents per month over the period considered.

An examination of Figure 14.4 shows vividly the declining trend in the Brent/Ekofisk price differential from $1.50 in mid-1980 down to 21–25 cents in 1984. In 1980–81 the fluctuations of this price differential were very marked, with peculiar reversal of signs (Ekofisk becoming cheaper than Brent on some occasions). This behaviour is one effect among many of the price disturbances on the world oil market caused by the Iranian Revolution and its aftermath. In 1983–4 the fluctuations in the spot price differential become less marked and the range of variations seems to be contained between 0 and 40 cents.

Rather surprisingly, the Brent/Flotta price differential shows no trend when the period January 1980–January 1985 is taken as a whole. This is due to a structural discontinuity between the sub-periods January 1980–June 1981 and June 1981–January 1985. There is a sudden surge in the value of the differential in mid-1981, meaning that Flotta, at that time, became dearer than Brent on the spot market. After this surge the price differential began to narrow from a peak of $2.50 in mid-1981 to an average of 25 cents in 1984.

A regression model for the period June 1981–February 1985 indicates a rate of decline in the Brent/Flotta spot differential equivalent to an average drop of 2.8 cents per month.

14.5   North Sea and Other Crude Prices

So far, the price movements and price differentials of North Sea crudes have been examined. North Sea crudes, however, are traded internationally and it is interesting to analyse the various relationships between the price of North Sea crudes and those of other world crudes.

We shall select for this purpose eight major crudes: the three North Sea varieties analysed above (Brent, Ekofisk, Flotta), the OPEC marker (Arabian Light), and other representative crudes from Nigeria (Forcados), the USA (WTI), the Gulf (Arabian Heavy) and the Soviet Union (Urals). The first step is to examine the correlation between price movements of these different crudes.

The close correlation between spot price movements of the three North Sea crudes has already been noted. Similar regressions – Brent spot price on the spot price of another crude without a constant – show that spot price movements in Brent are strongly correlated with Forcados, Arabian Light and Urals but that the correlation with WTI is weaker. The R^2 are 0.969, 0.942 and 0.939 for Forcados, Arabian Light and Urals respectively, 0.830 for Arabian Heavy but only 0.823 for WTI.

This may seem, at first, a slightly surprising result considering the keenness with which Brent and WTI traders watch each other's moves. In fact the price link between Brent and WTI does not operate through spot markets, for the simple reason that Brent must travel to reach the USA while WTI is available *in situ*. The price link is between forward Brent and WTI futures. Furthermore, the WTI market tends to be influenced in the short term by a number of domestic factors (some of them related to the behaviour of future prices and some to local conditions in the US oil industry), which do not always spill over onto the international petroleum market. It is therefore understandable that spot price relationships should be closer between export crudes that mutually influence each other in the European market, where they are freely tradable, than between them and a non-exportable domestic crude such as WTI.

14.6   Price Differentials : Brent and World Crudes

Price differentials between Brent, taken as representative of North Sea crudes, and five other varieties (WTI, Forcados, Urals, Arabian Light and Arabian Heavy) have moved in dissimilar ways during the period considered.

The statistical techniques used in section 14.4 to assess movements in the price differentials of North Sea crudes are also applied here. The first step is to test for any trend in price differential movements over time. The results are as follows:

— There is a statistically significant trend over the whole period in the Brent/Urals and Brent/Arabian Heavy price differentials. In both cases the trend implies a narrowing of the price differential. The reduction in the differential measured in absolute terms was equivalent to an average fall of 1.1 cents per month in the case of Urals (January 1980–February 1985) and 2.8 cents per month in the case of Arabian Heavy (July 1980–January 1985).
— There is no statistically significant trend over the period in the Brent/WTI, Brent/Forcados and Brent/Arabian Light price differentials.

Amalgamating these results with those on the price differentials of North Sea crudes reveals a clear pattern. Crude oil price differentials between extra-light (Ekofisk) and heavy crudes (Urals, Arabian Heavy) on the one hand and a conventional 'light' reference crude such as Arabian Light or Brent on the other have tended to narrow significantly. However, the price differentials of a fairly wide range of crudes (Brent, WTI, Forcados but also Arabian Light) with broadly similar properties did not display any notable tendency. In short, the narrowing of price differentials is a significant phenomenon insofar as very dissimilar crudes are concerned. The important implication of this finding is that attempts to fine-tune price differentials of crude varieties belonging to the same or to neighbouring clusters (see Chapter 2) can prove futile.

Statistical estimates of the average value of differentials, taking into account the possible existence of structural breaks in the spot price data over the long period considered, are presented in Table 14.5 below.

Table 14.5: Statistical Estimates of the Average Value of Price Differentials. Various periods.

Brent/WTI. The sub-period March 1983–November 1984 is identified as being fairly homogeneous. The average price differential is estimated at 71 cents during this sub-period with a standard error of estimate of only 26 cents. This means that the Brent spot price has tended to vary (with a 95 per cent probability) between $1.23 and 19 cents below the price of WTI. Outside this sub-period, particularly in December 1984/January 1985, but also in the first quarter of 1983, the Brent/WTI spot price differential became both negative and very large.

Brent/Arabian Light. There is no apparent or easily identifiable structural break in the period January 1980–January 1985. The spot price differential displayed considerable fluctuation throughout the period but the movement appears to have centred around an average of 98 cents. The standard error of estimate is high relative to the mean value (95 cents). This implies a range of variations from as much as $2.88 at one extreme to −$0.92 at the other during the period considered.

Brent/Arabian Heavy. The period considered is July 1980–January 1985. The estimate suggests that the Brent spot price was on average $3.07 above Arabian Heavy.

Brent/Forcados. The results are for the sub-period December 1982–January 1985. The estimated average value of the price differential in that period is 5 cents, which turns out not to be (from a statistical point of view) significantly different from zero. It can therefore be said that the spot prices of Forcados and Brent were on average equal over the period December 1982–January 1985.

Brent/Urals. The relevant sub-period is February 1983–February 1985. The Brent spot price, on average, exceeds that of Urals by only 21 cents. The standard error of estimate is 64 cents. Thus the Brent/Urals price differential tended to fluctuate, with a 95 per cent probability, between $1.49 and −$1.07.

Note: See Figures 14.5 to 14.10 at the end of the chapter.

These statistical estimates of average price differentials over a historical period may not have much relevance for future developments because some key differentials have tended to narrow significantly and are today much smaller than past 'averages'.

The more interesting observation is that the price differentials themselves are very volatile in the short term, showing great variability in their values. This is apparent from the figures, and can be show with greater precision by comparing the standard error of estimates with the mean value of the relevant price differential (see Table 14.6). The short-term volatility is so considerable that price differentials not infrequently change signs (i.e. a crude, such as Arabian Light, whose price is normally lower than Brent, may on occasions have a higher spot price than Brent).

Table 14.6: Variability of Price Differentials of Selected Pairs of Crudes

Crudes	Mean Value of Price Differential $	SEE $	Range of Variation of the Price Differential $	
Brent/Arabian Light	0.98	0.95	−0.92 to	2.88
Brent/Arabian Heavy	3.07	1.51	0.05 to	6.09
Brent/WTI	−0.71	0.26	−1.23 to	−0.19
Brent/Forcados	−0.05	0.46	−0.97 to	0.87
Brent/Urals	0.21	0.64	−1.07 to	1.49

Note: The minus sign means that the price of Brent is lower than that of the paired crude.

Such variability in the value of price differentials is both the effect and the cause of short-term trading on spot markets. The supply and demand schedules for different crude varieties shift continually relative to each other as a result of a multitude of short-term factors. These temporary disequilibria elicit spot price responses which, given time, are expected to bring the system to a new equilibrium. However, as the system begins to adjust, further changes in short-term circumstances create subsequent disequilibria before the system has had time to adjust, and prices respond again, moving either in the same direction as before or in the opposite direction. Thus small changes in the spot price of crudes can result in relatively large changes in the value of differentials (because differentials are much smaller numbers than the prices themselves).

The important point is that the equilibrium of the market is continually upset over very short periods of time, while the adjustments produced by price changes, however rapid these may be, do not obtain immediately. This means that the system is always in a state of disequilibrium. This situation creates opportunities for trading, partly because participants' views on the likely course of adjustments tend to diverge, and partly because differences in their own circumstances make a given change in the relative price of crudes more attractive to some than to others. Furthermore, when price differentials are volatile, end-users want to retain maximum freedom to shop around for the best buys, and maximum flexibility. This is helped by modern refinery technology, which enables them to switch easily from one crude to another, and new communication systems, which enable traders to track the market from hour to hour. Both these factors encourage short-term trading and rapid moves in response to changing price differentials, which in turn cause these differentials to vary once again.

14.7 Further Comparisons and Relationships between Price Behaviour in the North Sea and Other World Markets

We have so far examined a number of relationships between the prices of Brent and other crudes. These were:

— the correlation between spot price movements of Brent and seven other crudes
— the behaviour of price differentials between Brent and other major crudes

In order to gain a better understanding of the links between the North Sea market (as represented by Brent) and other oil markets we need to examine other relationships. These are:

— First, the nature of the link between the Brent price and oil product prices.
— Secondly, the relative variability of spot Brent prices and other crude spot prices.

(a) Brent Spot Prices and the Product Price Netback. The linkage between crude and product markets can be examined by analysing the relationship between the spot price of Brent and the product price netback for Brent (based on complex refining at Rotterdam).

The analysis, described in detail in Annex 6, looks at the relationship between the two series for the period January 1981–February 1985.

The margin was clearly positive for the period but it showed great variability across different months, and furthermore the margin declined from February 1983 onwards. Following earlier work done at the Institute on this subject, we divided the period into two sets of data:[1]

— those months for which the spot price was greater than the BNOC term price
— those months for which the spot price was less than the BNOC term price

The netback was regressed first on the spot price and secondly on both the spot and the term prices. The results show that a significant relationship exists between the spot price and the netback whether or not the spot is above the term price.

The results are interesting in some respects. The netback margin for Brent was found to be positive. There is, however, a big reduction in the value of the margin between the period before February 1983 and the period after. Statistical analysis suggested that some other factor, not reflected in spot prices, was systematically affecting netbacks. One possible explanation is that term prices were moderating the relationship between spot crude and netbacks in periods when the spot was below the term price. The term price exercised an influence because a certain proportion of North Sea crudes used by refiners was effectively purchased at the BNOC price, so that the average input cost of crude for a refiner was a composite of term and spot prices. An alternative explanation might be that in periods of excess supply, characterized by spot prices being below term, the margin is squeezed by falling product prices.

(b) The Short-run Variability of the Brent Price Compared with Other Prices. The comparison is made between the short-term variability of spot prices in six markets: Brent, WTI, Arabian Light, Forcados, Ninian and Dubai, using daily data over a relatively short period from January 1984 to February 1985.

---

[1] Bacon, R.W. *A Study of the Relationship between Spot Product Prices and Spot Crude Prices*, Working Paper, WPM5, Oxford Institute for Energy Studies, 1984.

Tests on the data, described in Annex 7, suggested that within this period there had been three distinct episodes. The first was a period of 'higher' prices until 20th June 1984; the second of 'medium' prices until 13th November 1984; and the last of 'lower' prices until end February 1985.

For each period the average variability was calculated by finding the standard error around the sub-period mean.

The results are quite strong. In all periods, Dubai and Arabian Light spot prices have the lowest variability (around 20–30 cents average variation a day); Forcados is in between the Gulf and the North Sea/US crudes (around 45 cents average variation a day); Brent, Ninian and WTI show high variability (with around 60 cents average variation a day).

In both the first and the third period WTI spot prices appear to have been more variable than North Sea crude prices. For example, between November 1984 and February 1985 WTI prices varied on average by 94 cents a day, North Sea prices by about 68 cents, and Arabian Light by only 22 cents.

Differences in spot price variability between crudes seem to be due to institutional factors rather than to the relative amounts of activity in the different markets. If the volume of transactions and the multiplicity of participants had a role to play we would have expected to observe the opposite pattern – with thin markets, such as Arabian Light, showing more variability than very active markets, such as WTI or Brent. The role of institutional factors is illustrated by two extremes; the futures market in the USA and the OPEC system.

— First, the existence of a futures market in WTI. Futures prices are very volatile because participants in the exchange respond very quickly and fairly sharply to changes in perceptions and circumstances. The WTI futures market naturally has a more direct influence on the WTI spot price than on other spot markets, and variability is probably transmitted from WTI futures to WTI spot.

— Secondly, the existence of an official price structure backed by some form of supply regulation in the OPEC countries. This factor has a strong stabilizing effect on the spot price of Arabian Light and, albeit to a slightly lesser extent, on Dubai. The influence is less marked on Forcados, as one might expect, but remains significant. The volatility of the spot price of Nigerian crudes, despite their strong Atlantic Basin links with Brent and WTI, is much less marked than for these latter crudes.

The high volatility of the prices of North Sea crudes may thus be explained by the absence of output regulation. However, there was a residual influence from the BNOC term price structure (operating through changes in the volume of oil traded on the spot market) which may have slightly reduced the volatility of North Sea crude prices compared with WTI.

14.8 Conclusions

The behaviour of prices and the price relationships analysed in this chapter point towards a general conclusion. The various world crude oil markets appear to be closely linked with each other, but each market has institutional characteristics of its own which may cause important differences in price behaviour. Furthermore, there are instances in which the relationships between markets break down for a while and prices appear to move independently in each market.

In all markets spot prices show variability, but volatility is greater when term or official price structures have the weaker influence. The relative variability is more marked for price differentials than for prices themselves. Crudes of different types compete with each other and spot markets are the place where buyers switch their custom from one crude to another and between different supply sources.

In a sense, the volatility of price differentials is the most important phenomenon. If the oil trading game is characterized by traders continually switching their dealings from one crude to another, it is on the movement of price differentials that the winners make their money and the less successful lose. Market structures such as futures and forward trading increase this variability by increasing the ease and frequency with which players can trade, taking advantage of short-run disequilibria without necessarily speeding up the fundamental role of adjustment in the oil market. Their development is therefore welcomed by the risk-taker and speculator.

In an oil world where excess supplies always threaten a vulnerable price level, volatility and short trading are both unavoidable and risky. No company can afford to refrain from playing the game that everybody else plays; all of them playing together can unwittingly bring about an outcome that nobody desires.

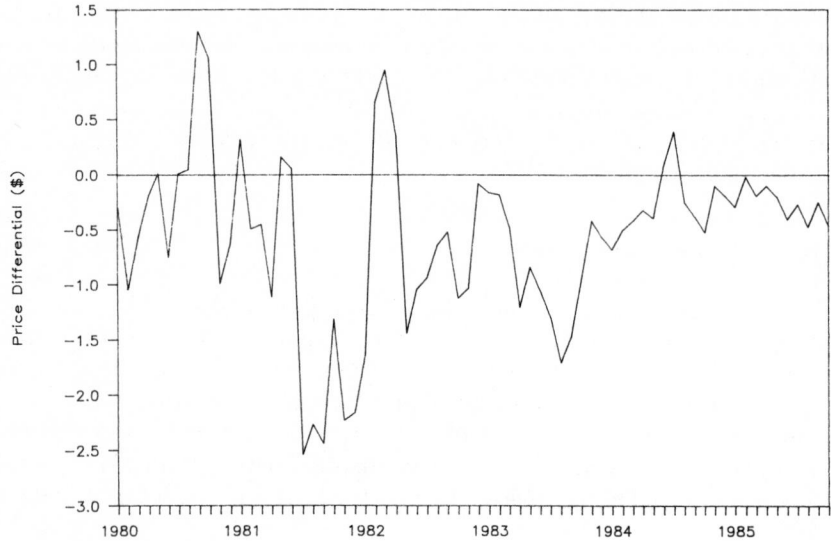

Figure 14.5   Flotta/Brent Differential. January 1980–October 1985

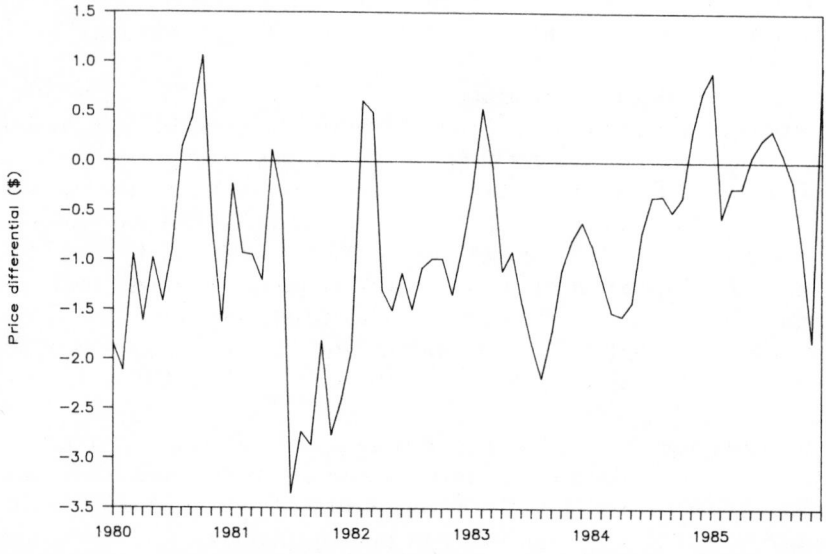

Figure 14.6   Arabian Light/Brent Differential. January 1980–December 1985

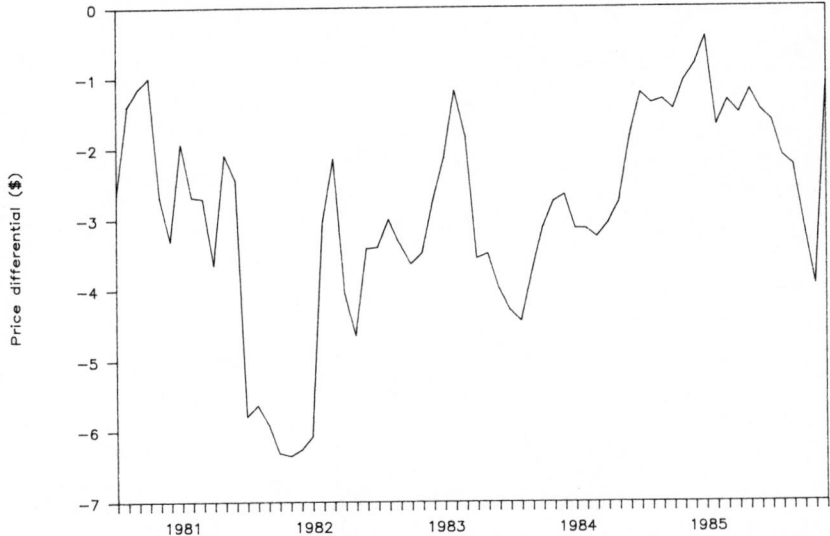

Figure 14.7 Arabian Heavy/Brent Differential. July 1980–December 1985

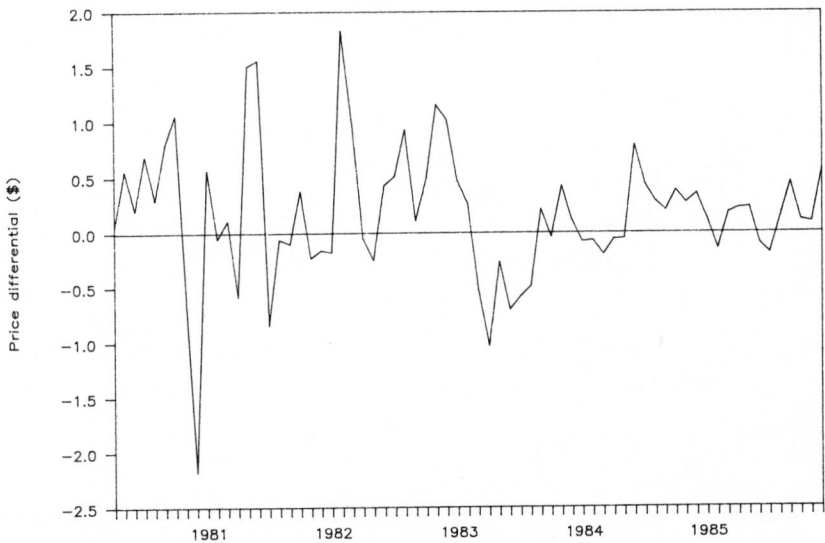

Figure 14.8 Forcados/Brent Differential. April 1980–December 1985

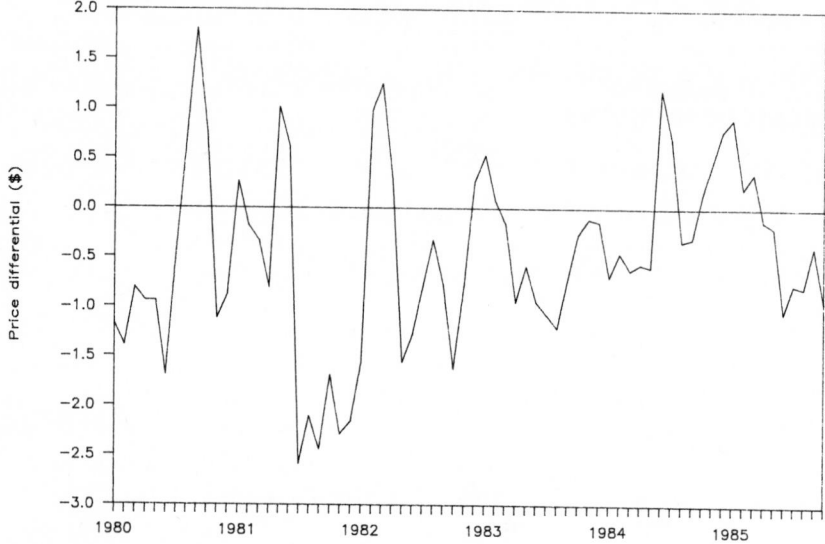

Figure 14.9   Urals/Brent Differential. January 1980–October 1985

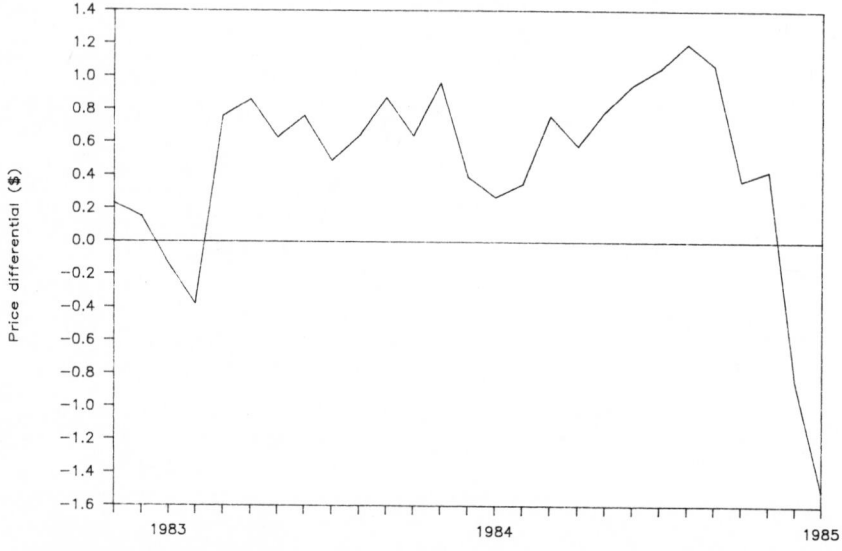

Figure 14.10   WTI/Brent Differential. November 1982–January 1985

CHAPTER 15

THE ROLE AND INFLUENCE OF THE NORTH SEA IN THE WORLD OIL MARKET

15.1 Introduction

In August 1984 three members of the Institute staff, each on his own, conceived the idea of a study of the market for North Sea crude oil. This meeting of minds arose out of a shared perception that the North Sea was playing an undefined, yet important, role in the world petroleum market. We recognised the need to do some intensive research to discover both the nature and the limits of this role.

When, nine months later, we reached the end of our research, we felt that a better understanding of these issues had been attained. The purpose of this chapter is to draw together the results of our research to provide the elements for a comprehensive description of the North Sea market in its international context, and to take a view on possible future developments.

15.2 The Centrality of the North Sea

The role and influence of the North Sea do not derive primarily from its size relative to other parts of the world oil industry. The USA and the Soviet Union dwarf the UK and Norway if oil output is taken as a criterion of size. The Middle East in general, and Saudi Arabia in particular, dwarf the North Sea if oil reserves are the relevant criterion. Yet size *is* a significant factor today. In 1985, largely as a result of a considerable reduction in the volume of OPEC production (a decline for which UK and Norwegian oil are responsible in some part), the North Sea moved up in the ranking of world producers and exporters to reach the third place in the oil output league (ahead of Saudi Arabia) and the second place in the oil export league (just behind Saudi Arabia).

The influence of the North Sea is attributable first to its *centrality* in the world petroleum market, and secondly to certain factors related to *size*. These two factors are related because centrality would not have obtained had the output of UK and Norwegian oil or the export volume been insignificant. However, centrality is essentially due to the development of particular market structures within which North Sea crudes are both continually and very actively traded in a visible, if not completely transparent, way. These market structures have enabled North Sea prices to perform a barometric role for the formation of oil prices elsewhere in the world.

Since size is a factor contributing to centrality, it is legitimate to take this concept as our starting point. North Sea crude production has grown rapidly in a period during which world demand for oil has been either stagnant or on the decline. To be sure, other non-OPEC production also increased in the first half of the 1980s; in this respect the North Sea is part of a wider supply phenomenon. The growth of North Sea oil output in a declining market thus had a significant impact on OPEC, as became apparent during the oil price crisis of early 1986.

Another aspect of size is the proportion of oil output that is traded. In OPEC, and in some non-OPEC countries such as Mexico and Egypt, the volume of oil retained for domestic use is disposed of internally by the country's national oil corporation; this volume is not traded on a market. Trade is limited to export crude. In contrast, a very large proportion of North Sea oil, whether used domestically by the UK or Norway or exported, is traded between oil companies (tax spinning has considerably reduced the portion that remains within the channels of integrated companies and bypasses the market). If traded oil is taken as the measure of size, the North Sea ranks second in 1985 after the USA.

A third aspect of size is the number of transactions that trade in oil generates. In this respect, WTI, despite its relatively small output volume, is a very significant crude because of the turnover of the New York Mercantile Exchange (Nymex). Similarly, transactions in the Brent market multiply the physical volume of spot market oil by an average coefficient of 10–15. This multiplier gives Brent a much higher profile on the world petroleum scene than would be warranted by its output volume (900,000 b/d), significant as this may be.

The centrality of the North Sea is enhanced, therefore, by the size of the place it occupies in the world of crude oil trading. Size, however, is an enabling factor, which does not of itself reveal very much about the central role played by the North Sea in the world petroleum market. Insight on this issue can only be gained by examining the functions and the behaviour of North

Sea markets, in other words the qualitative rather than the quantitative aspects.

The fundamental characteristic of the North Sea market is that it responds to shifts in demand and to outside shocks almost exclusively by price adjustments. This derives from the strategy of output maximization in the medium term (there may be short-term and temporary exceptions) followed by both companies and governments in the UK and Norway. When supply is a datum, at a given point in time, variations in demand can only be accommodated by price changes.

This is in marked contrast to the behaviour of markets for OPEC oil, in which demand shifts and other shocks are largely (though not exclusively) accommodated by quantity adjustments.

Markets that adjust through price changes continually transmit signals; the higher the profile of these markets, and the greater the transparency of their operations, the better is the transmission of signals to their own participants and to other parts of the oil world.

Though far from perfect, the transparency of the Brent price (and to a lesser extent of other North Sea crude prices) is sufficient for a swift and fairly clear transmission of signals. Because the North Sea market, as argued above, is large, and because it is transparent, variations in its prices put pressure on other markets. The WTI futures market performs a similar role (in fact the transparency is much greater) but its impact is weakened by the absence of a physical interface between WTI and other crudes outside the USA.

Furthermore, North Sea crudes have recently gained in leverage, not only because of large traded volumes and significant penetration in two of the three largest consuming areas in the world (North West Europe and the USA but not Japan), but also because of the convergence of price differentials towards a middle-range cluster of highly sought-after, good quality, all-purpose crudes in which North Sea varieties figure prominently.

The centrality of the North Sea market, resulting from size in the various senses distinguished above, from the price adjustment characteristics of its mode of operation, from price transparency and from the centrality of the crudes themselves in the crude oil supply-mix, enables it to play a barometric role. Its price signals are not only transmitted but also widely received, and often interpreted as indicators of changes in supply/demand conditions and in the behaviour of important economic agents such as governments and large oil companies.

15.3 The Barometric Role of Prices

The significance of this barometric role is better understood when related to a simple model of the oil market. We have already suggested that the world petroleum market consists of two parts:

— One in which changes in demand elicit only price adjustments because supplies are price inelastic in the short term; this part includes the USA, the North Sea, other non-OPEC areas and even some OPEC countries.
— Another in which changes in demand will normally elicit quantity adjustments as if supplies were totally price elastic; this part is represented by the hard core of OPEC.

These two parts or regions are linked to each other because petroleum is in effect a set of highly traded commodities, each of which is a very close substitute for the others. (Situations in which different segments of the world petroleum market cease to relate to each other occur on occasion but may be of short duration.) In a normal state of affairs the linkage between the two parts of the market operates very simply. The emergence of, say, excess demand in the price responsive region immediately pushes up prices in this region. This reaction diverts some demand to the OPEC part of the market and is met by a commensurate increase in supplies at the given price (this is a pure quantity adjustment). As demand is diverted, prices in the first region fall back towards their original level and the initial disturbance is eventually eliminated. Subsequent disturbances, whether they follow or superimpose themselves on the previous one, have similar effects. In this simple case, the stability of the system is not upset by price variations in the first region, and price movements do not play a barometric role because OPEC, without much ado, absorbs all shifts in demand through a change in its output.

However, situations arise when OPEC is unable to absorb the demand shifts. This can happen in two different instances:

— The first, illustrated by the events of 1979–80, is a situation in which OPEC is producing at full capacity. In this case an *increase* in demand for oil cannot be accommodated by a quantity adjustment and the OPEC price is raised. (Note that in this interpretation price increases by OPEC reflect a failure of the adjustment process and not a successful act of cartelization.)
— The second instance, illustrated by the period 1982–5, is a

situation in which OPEC is producing much below capacity. In this case the continuing *decrease* in the demand for oil may cause some OPEC members to respond by unofficially lowering their prices. There is reluctance to let the system adjust through a pure quantity change because members attempt to protect their share of a shrinking market or their share of revenues. Whenever unofficial price discounting turns into open price competition, OPEC meets in crisis and attempts to resolve the problem temporarily by a downward adjustment of official prices (e.g. in March 1983 when the marker price was reduced by $5, and in January 1985 when it was reduced by $1).

In these two situations, price signals emanating from the non-OPEC part of the market (the region that always adjusts through prices alone) play a barometric role. Under these circumstances the OPEC region is unable to absorb all demand shifts through pure quantity adjustments and needs to vary prices from time to time.

Prices in the non-OPEC region rise on *trend* when the market is in a medium-term state of overall tightness (as in 1979–80) and will fall on *trend* when the market is in a general state of slackness (as from 1982 to date). Prices, however, show very marked *variability* in the short and the very short terms around these trends. A medium-term rise on trend will involve a very large number of instances in which prices fall over a sequence of days, if not longer. Similarly, a medium-term fall on trend will involve prices going up as well as down in the shorter period.

Because the OPEC region normally adjusts through quantity and only reluctantly through prices, it will not respond to their variability in the barometric role by closely following all the up and down moves. The response in the OPEC region to these barometric price changes tends to be asymmetrical.

In today's conditions of slack demand, the OPEC response to a short-period price rise, which goes against the medium-term trend, is not to increase prices but to put more oil on sale; but the response to a short-period price fall, which further emphasises the medium-term trend, may be an informal lowering of prices by some OPEC countries. As the number or the frequency of these informal 'price adjustments' increases, OPEC is obliged to reduce its official prices.

The significance of price variability in the non-OPEC part of the world market can now be better gauged. The movement of a barometric price that is very variable involves very frequent and sometimes sharp ups and downs. Because of the asymmetry in responses, the 'ups', however frequent, have no price impact; but the more frequent the occurrence of 'downs' the greater the chance of a slide in the OPEC price structure.

15.4 Market Stability

In a world in which the relationship between oil demand and potential supplies is extremely unbalanced, markets left to their own devices would inevitably bring down prices, in the short and medium terms, to abysmal levels. So far this has not happened because a significant, albeit continually shrinking, part of the market has tended to absorb the imbalance by shutting in capacity rather than lowering prices.

It is, of course, the fundamental supply and demand imbalance which in the end brings about the price changes.

The objective of exporting countries in the face of supply/demand imbalances simply consists of making this period of 'no price adjustment' as long as possible. The purpose is to gain time, which in effect means to gain revenues, while awaiting a favourable future upturn of the market cycle.

However, the duration of the period over which the change can be resisted can be shortened by the destabilizing impact of a volatile barometric price. North Sea prices are both barometric and volatile. We have also seen that, in the North Sea, the Brent market involves much speculative activity. When prices fall, for whatever reason, some speculators begin to think that the OPEC part of the market will not be able to absorb the associated reduction in demand for long. They speculate that prices will fall further and that end-users will curtail their purchases for as long as they can, thus bringing about a further fall in a self-fulfilling prophecy.

Those who observe price movements minute by minute, or day by day, without the benefit of distance, cannot easily distinguish those changes caused by temporary (sometimes trivial) factors from those changes related to more fundamental forces. This tends to confuse the responses and may on occasion prove to be another destabilizing factor.

The upshot of this analysis is that the North Sea is not the primary cause of market imbalances (although the contribution it makes to the fundamental imbalance through the growth of its supplies is not negligible), but it acts as a highly sensitive and dynamic transmitter of potentially destabilizing price signals.

In this respect local changes in the conditions of North Sea markets, as occurred in mid-1984, tend to have much wider repercussions than their significance warrants.

The findings of this study reveal a remarkable coincidence of phenomena in the period starting in July 1984. There was a very marked dip in the spot prices of North Sea crudes accompanied by significant increases in the volume of tax-spinning oil, in the volume of BNOC's supplies to the spot market, and in the level of activity on the Brent market. There was also a rise in the volatility of North Sea spot prices which reversed an earlier trend of reduced variability.

These disturbances were transmitted to the OPEC part of the market when BNOC decided to change its term price in October 1984. It was at this juncture that an important aspect of the role of BNOC became clear. So long as BNOC operated and was able to defend a term price structure, the transmission of price signals from North Sea spot markets was considerably dampened. In other words BNOC acted as a first line of defence. Price crises occurred whenever the movements of North Sea spot prices forced BNOC to change its term price. In these situations BNOC appeared to be responsible for the crises and got the blame. What was always forgotten was its success in delaying them in the meantime.

In the end, OPEC reduced its official prices by a mere dollar, but its main response was a further significant output adjustment. Spot prices, brought down in December 1984/January 1985 by expectations of an oil price collapse, rebounded for a while; in both cases spot price movements overshot, giving exaggerated indications first of excess supply and then of excess demand.

At the same time BNOC was abolished. The abolition means that the transmission of spot price signals from the North Sea will be neither hindered nor delayed. Nothing stands in their way any longer.

The current oil price crisis, which began in mid-1985, was triggered by the pressure exerted from falling spot prices on a large number of OPEC countries that responded by giving their customers bigger price discounts. The burden of quantity adjustments initially fell on two countries, on Saudi Arabia in a conspicuous way and on Mexico in a less publicized but more painful manner. Mexico yielded first by lowering its price on 11th July 1985; Saudi Arabia followed in September by entering into 'netback' sales agreement with oil companies. It abandoned, for the time being at least, its attempts to stabilize prices through output reductions. Saudi Arabia decided to secure a fixed volume of oil exports and accept the adverse price consequences of this new policy.

15.5   The Future Role of the North Sea Market

As pointed out elsewhere, UK oil production most probably reached a medium-term peak in 1985 but Norway is expected to increase its output in the next few years by some 300,000 barrels. What is therefore certain is that the significant growth of North Sea oil production in recent years will not continue at the same rate. The growth aspect of the pressure exercised on the world petroleum market by the North Sea will undoubtedly recede.

The North Sea will however remain important in terms of its centrality to the market. In the most plausible scenario, North Sea markets will continue to be very active. In fact activity may become more diversified if the methods of forward trading developed on the Brent market are adopted fully for other crudes such as Forties, Ninian, Flotta and perhaps Ekofisk.

Forward trading allows both physical hedging and paper speculation. It is a natural stage in the evolution of the oil market, given that the traditional means of reducing uncertainty – the term contract – has now almost entirely disappeared. Forward trading has already been extended from Brent to Dubai and Soviet gas oil. It is just a matter of time before it becomes fairly general in the North Sea.

This extension takes the form of new market structures emerging for other crudes, following the pattern of Brent arrangements but with appropriate modifications (Brent has logistic features that make it easier to trade forward than, say, a North Sea offshore loading crude or an OPEC crude not handled for that purpose by the original producer).

Though we can see expansion taking place laterally, and indeed we expect this development, we are less sure about the room left for a quantitative increase in the Brent market itself. The issues are as follows.

First, the recent growth in Brent market activity, such that levels of trading at end 1985 almost certainly exceeded those that might have been expected even a few months previously, has imposed a strain on the operation of the market. Clearing the market in a given month involves the co-operation of all the participants, and the larger the number of deals and the number of participants the more complex the task becomes. Recent innovations, such as book-outs, undoubtedly help, but are not sufficient. So far the market has survived on the goodwill of the large primary sellers and end-users of Brent, mainly major oil companies; but even this goodwill cannot always avert failures, as recent defaults have shown.

Secondly, growth can be unsustainable in economic terms. If our analysis of the role of hedgers and speculators (as defined in this study) is correct, it would appear that, for a given range of differing expectations, there is an upper limit to the ratio of paper trading to wet deals in the long run. Other things being equal, the volume of Brent production determines this upper limit.

Thirdly, expansion resulting from cross-trading other North Sea grades against Brent would over-extend the Brent market in the short run unless other North Sea crudes develop similar market structures or the Brent contract is modified to include other grades. To understand this difficulty we must return to our analytical distinction between hedging and speculating (see section 13.9). When companies cross-trade they regard this activity as hedging, and this suggests that they are adding to the 'kitty' for speculators and that activity can thus safely increase. In our model cross-trading is not hedging (because the company ends up by closing out its Brent position) but speculating; it does not add to the 'kitty' but competes for the limited amount available in it.

Fourthly, the market in Brent may be kept buoyant in the short run by some companies losing overall on the speculative aspects of their trading. These losses in effect increase the size of the kitty and permit a high number of deals. As such companies come to reassess their performance, they will either trade less (only on more certain profit opportunities) or withdraw from the market altogether. As the learning phase passes, the number of deals may tend to reduce.

Finally, it should be recalled that at any one time the liquid base of the Brent market is not the whole of Brent production but the quantities put in by primary sellers. These quantities exclude volumes sold on term contracts (now very rare) and volumes retained by integrated oil companies for their own use. The latter depend on the extent of tax spinning. The abolition of BNOC, followed by clarifications made by the OTO about the valuation method for tax purposes of oil transferred internally, may significantly reduce the motivation for tax spinning. However, there are arguments, as stated in section 8.10, which suggest that some tax spinning could continue despite these changes in circumstances. It is difficult to tell; but a reduction in tax spinning not matched by increases in spot market supplies arising for other reasons would reduce the sustainable limit of paper activity on the Brent market.

To sum up; our view of future developments expects the development of market structures, broadly similar to the Brent model, elsewhere in the North Sea and perhaps in other parts of the world. Spot and forward trading are here to stay, and will inevitably expand unless OPEC adopts a novel strategy, reintroducing long-term contracts for the bulk of its supplies. The probability of such an innovation is not very high.

The increase in the volume of forward trading can only take place by diversification into crudes other than Brent. The Brent market as it operates today is probably saturated, and any further expansion within its confines would cause increasing strains. The major participants have already perceived the need for greater formality and regulation, and for innovative changes in procedures. They have the power and the means to impose them; we know that the Brent market owes much to their interest in its survival and to their goodwill.

PART V

TECHNICAL ANNEXES

ANNEX 1

TEMPORAL PRODUCTION PATTERNS FOR THE UKCS AND NORWEGIAN CONTINENTAL SHELF

A1.1 Introduction

The purpose of this annex is to provide a statistical analysis of temporal production patterns for the UKCS and Norwegian Continental Shelf over the period January 1980–February 1985. Statistical techniques are used to determine trend growth rates and to test for seasonal variations in production. The oil production data are taken from the statistics published by Wood Mackenzie. These enable us to construct a monthly series from January 1980 to February 1985. The data sample, therefore, contains sixty-two observations.

A1.2 Temporal Production Patterns: UKCS

(a) *Total UKCS*. In order to test for growth and seasonality we regress total UKCS oil production (NSO) on a linear time trend and a set of quarterly dummies. The results are presented in Table A1.1. The dummy variables are designed to represent the winter, spring and summer quarters so that the equation is normalized on the autumn quarter. However, there are a number of possible interpretations as to the appropriate months for each quarter. Hence, we experiment with three versions of each quarter so that there are three sets of quarterly dummies.

In Table A1.1 the coefficients on the TREND variable are always significant and indicate a trend increase in UKCS oil production of approximately 19,500 b/d per month over the data period.[1] In addition, there is strong evidence of seasonality in UKCS oil production. For example, in equation 2, the significant coefficient on the DW2 variable implies that the winter quarter is associated with a rise in production of 60,500 b/d relative to the autumn quarter. In contrast, the summer quarter sees an output fall of 61,500 b/d relative to the autumn quarter. The spring quarter is also significant in two of the three regressions but its change of sign suggests that its significance is largely due to the fringe months of February and June.

(b) *UKCS Oilfields*. The same analysis is now undertaken for the four major UKCS oilfields – Brent, Forties, Ninian and Piper – and for the Rest of the UKCS. In addition, some evidence of output seasonality for the Rest of the UKCS leads to an

---

[1] i.e. the average daily output for a month is 19,500 barrels greater than the average daily output from the previous month.

Table A1.1: Temporal Production Patterns for the Total UKCS (January 1980–February 1985)

Dependent Variable	Constant	TREND	DW1	DSP1	DSM1	DW2	DSP2	DSM2	DW3	DSP3	DSM3	R^2	SEE	DWS
1. NSO	1422.0* (26.4)	19.6* (0.5)	81.6* (27.2)	91.9* (27.2)	1.0 (27.6)							0.96	75.6	1.40
2. NSO	1460.7* (26.5)	19.5* (0.5)				60.5* (26.4)	25.1 (27.4)	−61.5* (27.3)				0.96	74.6	1.25
3. NSO	1487.5* (27.0)	19.5* (0.5)							38.4 (26.7)	−48.8** (27.6)	−73.8* (27.5)	0.96	75.2	1.40

Notes: * denotes significance at 5 per cent confidence level
        ** denotes significance at 10 per cent confidence level
        Standard errors in parentheses

Dummy variables 1:   DW1 = 1 for November, December, January; = 0 otherwise
                     DSP1 = 1 for February, March, April; = 0 otherwise
                     DSM1 = 1 for May, June, July; = 0 otherwise

Dummy variables 2:   DW2 = 1 for December, January, February; = 0 otherwise
                     DSP2 = 1 for March, April, May; = 0 otherwise
                     DSM2 = 1 for June, July, August; = 0 otherwise

Dummy variables 3:   DW3 = 1 for January, February, March; = 0 otherwise
                     DSP3 = 1 for April, May, June; = 0 otherwise
                     DSM3 = 1 for July, August, September; = 0 otherwise

examination of production patterns in the Fulmar, Claymore, Dunlin and Cormorant oilfields. An identical set of regressions has been estimated but we report only the preferred regression for each field. The results are given in Table A1.2.

The results on production patterns in the Brent field are shown by equation 1 of Table A1.2. The TREND variable indicates a trend increase of 6000 b/d per month over the data period. The estimates confirm the seasonal nature of Brent oil production. The significant coefficients on DSP3 and DSM3 show that the spring and summer quarters are associated with production falls of 51,500 and 41,400 b/d respectively, relative to autumn. Equation 2 of Table A1.2 shows that the Forties field has experienced a significant trend decrease in production of approximately 1700 b/d per month. The Forties field also displays strong seasonality in its production patterns. The winter and summer dummy variables are significant. Output rises by 18,700 b/d in the winter and falls by 33,100 b/d in the summer relative to the autumn quarter.

The results for the Ninian field (equation 3) suggest that there is no regular temporal pattern of production. The TREND and quarterly dummy variables are all highly insignificant. However, a plot of Ninian oil production distinguishes two sub-periods. From January 1980 to June 1982 production shows a steady rise which is followed by a clear downward trend to the end of the data period (February 1985). The data sample is divided, therefore, into the two sub-periods in order to identify the distinctive trend for each period. The results are presented in Table A1.3.

The significant coefficients on the TREND variable in both equations confirm the increase in production over the first half of the data period and the decline over the second half. Oil production from the Ninian field shows a trend increase of 3900 b/d per month between January 1980 and June 1982. In contrast, there is a trend decrease in output of 3100 b/d per month over the period July 1982–February 1985. In both equations the seasonal dummies remain insignificant.

For the Piper field (see equation 4 of Table A1.2), the significant coefficient on the TREND variable indicates that there is a slight downward trend over the whole period of 600 b/d per month. The only significant dummy variable is DSP3 (April, May, June) and this probably reflects the special circumstances of the Piper field during the period April–June 1984.

Production data for the Rest of the UKCS is simply obtained by taking the difference between total UKCS production and the combined production of the four major UKCS fields (Brent, Forties, Ninian and Piper). The estimates for the Rest of the UKCS are given by equation 5 of Table A1.2. The significant coefficient on the TREND variable shows a strong upward trend of 15,800 b/d per month. The performance of the quarterly dummies implies that only the spring quarter displays any significant variation relative to the autumn quarter. Even in this case the coefficient is only significant at the 10 per cent confidence level. Nevertheless, it suggests that output rises by 47,800 b/d in the spring quarter relative to the autumn quarter.

The data period for the Fulmar field is shorter because oil production only began in February 1982. The results for the Fulmar field show a significant TREND

Table A1.2: Temporal Production Patterns for the Major UKCS Oilfields (January 1980–February 1985)

Dependent Variable	Constant	TREND	DW1	DSP1	DSM1	DW2	DSP2	DSM2	DW3	DSP3	DSM3	R^2	SEE	DWS
1. BRENT	142.2* (22.0)	6.0* (0.4)							2.6 (21.8)	−51.5* (22.5)	−41.4** (22.4)	0.77	61.3	0.55
2. FORTIES	514.2* (10.9)	−1.7* (0.2)							18.7** (10.8)	−11.2 (11.2)	−33.1* (11.1)	0.57	30.4	1.86
3. NINIAN	274.0* (13.4)	−0.1 (0.3)							−7.8 (13.2)	−1.0 (13.7)	−1.3 (13.6)	0.01	37.1	0.33
4. PIPER	220.5* (6.2)	−0.6* (0.1)							0.7 (6.2)	−15.1* (6.4)	−1.1 (6.3)	0.27	17.3	1.33
5. REST	328.8* (26.7)	15.8* (0.5)				30.3 (26.6)	47.8** (27.6)	15.1 (27.5)				0.93	75.2	0.43
6. FULMAR	−21.6 (24.4)	2.6* (0.5)	27.0** (14.7)	18.2 (14.3)	7.1 (14.7)							0.47	30.9	1.37
7. CLAYMORE	88.8* (2.3)	0.2* (0.05)	−2.6 (2.4)	−2.5 (2.4)	−1.9 (2.5)							0.17	6.7	0.74
8. DUNLIN	99.2* (3.5)	−0.9* (0.1)				14.5* (3.5)	14.2* (3.6)	19.1* (3.6)				0.77	9.8	1.50
9. CORMORANT	14.6* (3.8)	0.4* (0.1)							4.1 (3.8)	−8.7* (3.9)	−2.6 (3.9)	0.36	10.6	1.09

Notes: The Fulmar oil production series runs from February 1982 to February 1985.
REST is defined as total UKCS oil production *less* production from the Brent, Forties, Ninian and Piper fields.

Table A1.3: Temporal Production Patterns for the Ninian Oilfield

Sub-period	Constant	TREND	DW3	DSP3	DSM3	R^2	SEE	DWS
1.	215.6*	3.9*	−3.6	−10.7	−4.4	0.78	17.9	0.69
	(9.8)	(0.4)	(9.5)	(9.4)	(10.4)			
2.	410.1*	−3.1*	1.0	6.1	−0.3	0.66	20.3	1.70
	(19.7)	(0.4)	(9.9)	(10.7)	(9.6)			

Notes: Sub-period 1 runs from January 1980 to June 1982.
Sub-period 2 runs from July 1982 to February 1985.

variable (equation 6 of Table A1.2). Its coefficient indicates a trend increase of 2600 b/d per month over the period February 1982–February 1985. There is some weak evidence of output seasonality. The coefficient on the winter dummy variable, DW1, is significant at the 10 per cent confidence level and indicates an output rise of 27,000 b/d in the winter quarter relative to the autumn quarter. The other dummy variables are all insignificant.

The results for Claymore oil production (equation 7) also show a significant upward trend. However, this is very slight – around 200 b/d per month over the data period. There is no evidence for production seasonality since all the quarterly dummies are insignificant.

In equation 8 of Table A1.2, the significant coefficient on the TREND variable indicates an average trend decrease in output from the Dunlin field of 900 b/d per month. Furthermore, the three quarterly dummies are all significant with positive coefficients. These coefficients show monthly production rises of 14,500, 14,200, and 19,100 b/d in the winter, spring and summer quarters respectively, relative to the autumn quarter (October, November, December).

It should be noted that production from the Cormorant field was not continuous over the data period. There was no production in the Cormorant field during the period May–September 1981. Moreover, no production was recorded for October 1980 or April 1982. Production in the Cormorant field shows a trend increase of 400 b/d per month (equation 9). There is also some evidence of seasonality. The spring dummy variable, DSP3, has a significant negative coefficient which indicates a fall in output of 8700 b/d in the spring quarter relative to the autumn quarter. Such a result may largely reflect the shut-down during the period May–September 1981.

A1.3 UKCS Oil Production and Prices

(a) *UKCS Oilfields*. It is possible that the marked fluctuations in production of UKCS oilfields are related to economic considerations (rather than weather/maintenance-related factors). A check on this is to relate production to the spot price of oil. As a general indicator of prices we use the Brent monthly average spot price from *Petroleum Argus* (PRBREN) which is available from July 1982 to February 1985. We can now estimate a model in which production is related to a constant, a trend, the seasonal dummy variables and the spot crude price. The data sample covers the period July 1982–February 1985 (thirty-two observations).

The results for the four major UKCS oilfields – Brent, Forties, Ninian and Piper – are presented in Table A1.4. Of the smaller UKCS fields, only Dunlin's production shows evidence of price sensitivity. The results for Dunlin are included, therefore, in Table A1.4 while the estimates for the rest of the fields are not reported. Under the assumption that no individual field is large enough to influence the price, the coefficient on the price variable is expected to be positive. We again experiment with the three sets of quarterly dummies but only the preferred equation is reported for each field.

The results for the Brent field are shown by equation 1 of Table A1.4. Over the period July 1982–February 1985 the spot crude price is positively related to production as expected but it is not quite significant at the 10 per cent level. The significant TREND variable shows a smaller trend increase (3200 b/d per month) over the sub-period compared with that for the whole period (6000 b/d per month). The seasonality of production comes through much more strongly than before. Seasonal fluctuations in production appear to increase in 1983–4 relative to earlier years. The coefficients on the winter and spring dummy variables imply rises in output of 68,000 and 61,400 b/d respectively, relative to the autumn quarter. In contrast, the summer quarter (May, June, July) sees a fall of 86,600 b/d compared with the autumn quarter (August, September, October).

The estimates for the Forties field (equation 2) show the oil price to be highly insignificant. Moreover, over the sub-period the TREND variable becomes insignificant. As regards seasonality, the significant positive coefficients on the DW1, DSP1, and DSM1 variables indicate that output rises by 29,900, 31,900, and 43,400 b/d in the winter, spring and summer quarters respectively, relative to the autumn quarter. The previous significance of the DW3 and DSM3 dummies disappears over the shorter period.

There is no evidence of price sensitivity in the Ninian field. Equation 3 of Table A1.4 also confirms the absence of any significant seasonality in Ninian's production patterns. However, the data for the period July 1982–February 1985 manage to capture a significant downward trend of 3400 b/d per month.

The results for the Piper field (equation 4) indicate a significant *negative* association between oil production and prices. Furthermore, the significant coefficients on the TREND and DSP1 variables show a strong downward trend and some output seasonality. However, such results are probably contrived from the peculiarity of the shifts in the spring of 1984. There is so little variation generally in the Piper production series that all correlations must be questioned.

In equation 5 of Table A1.4, there is a significant *positive* association between Dunlin production and the spot crude price. In addition, the TREND and seasonal dummy variables perform even more strongly over the shorter data period. The Dunlin field now exhibits a trend decrease of 1100 b/d per month. The winter, spring and summer quarters show increases in production of 15,200, 14,900, and 19,200 b/d respectively, relative to the autumn quarter.

Table A1.4: The Relationship Between Production and Spot Crude Prices for UKCS Oilfields

Dependent Variable	Constant	PRBREN	TREND	DW1	DSP1	DSM1	DW2	DSP2	DSM2	R^2	SEE	DWS
1. BRENT	-75.5 (302.1)	11.0 (7.9)	3.2* (1.5)	68.0* (23.3)	61.4* (25.3)	-86.6* (23.9)				0.62	46.6	2.10
2. FORTIES	455.9* (200.3)	0.4 (5.2)	-1.3 (1.0)	29.9** (15.4)	31.9** (16.8)	43.4* (15.8)				0.21	30.9	2.03
3. NINIAN	486.9* (127.1)	-2.1 (3.3)	-3.4* (0.6)	4.8 (9.8)	-6.8 (10.7)	10.5 (10.0)				0.68	19.6	1.87
4. PIPER	489.3* (118.0)	-6.9* (3.1)	-1.8* (0.6)	-3.1 (9.1)	-26.3* (9.9)	-11.6 (9.3)				0.25	18.2	2.01
5. DUNLIN	100.9* (3.5)	0.3* (0.1)	-1.1* (0.1)				15.2* (3.4)	14.9* (3.5)	19.2* (3.5)	0.78	9.5	1.68

In summary, the individual UKCS fields show little evidence of a positive production response to crude prices over the data period. Consequently, as it is observed that seasonal output variations increased during 1983 and especially in 1984, a shorter data period is tried for the one *major* field that shows marginal price sensitivity (i.e. Brent). The Brent regressions are repeated for the data period January 1983–February 1985 (twenty-six observations) and the results are given in Table A1.5. The TREND and seasonal dummy variables perform very much as before. However, an interesting feature of the results is the performance of the price variable in equation 3. It is significant at the 10 per cent confidence level and has the expected positive coefficient. The size of the coefficient implies that a $1 change in the Brent spot price is associated with increased output of 21,200 b/d. Here then is some tentative evidence for a price-sensitive supply function for Brent crude oil.

Nevertheless, the evidence in support of price sensitivity is generally inconclusive. It is possible that the retention of the dummy variables in the equations may subdue the explanatory power of the spot crude price. Hence, some of the regressions are re-estimated without the seasonal dummy variables. The results are given in Table A1.6. The equations clearly suggest that there is no significant relationship between oil production and spot crude prices. In all three equations the coefficient on the PRBREN variable is insignificant.

Finally, we test the hypothesis that oil producers adjust output in response to variations in the value of the crude oil once converted into products, i.e. the netback. We estimate a model in which oil production is related to a constant, a trend and the Brent netback value. The Brent netback is taken as a general indicator of movements in spot product prices. The Brent netback series (for basic refining at Rotterdam) is provided by an industry source and runs from January 1980 to December 1983. The results are presented in Table A1.7.

The netback variable generally performs badly. It is significant only in the equation for the Forties field and even then only at the 10 per cent confidence level. We also have access to a Brent netback series which refers to *complex* refining at Rotterdam for the period January 1981–December 1983. Using this netback series the equations in Table A1.7 are re-estimated. In this case, the netback variable is insignificant in all three equations and the results are not reported.

(b) Total UKCS. There is a problem of causality underlying the relationship between UKCS production as a whole and prices. An increase in production could have a negative effect on the spot crude price. However, it is argued that the spillover from other crudes is likely to weaken this effect and that a positive supply response will be observed. The basic model is estimated for the total UKCS and the results are given by equation 1 of Table A1.8.

The data from July 1982 to February 1985 produce a positive but highly insignificant price effect. The seasonal dummy variables perform very much as before although the seasonality is somewhat more pronounced. The winter quarter is associated with an output rise of 80,500 b/d relative to the autumn quarter while the summer quarter shows a fall of 99,000 b/d. The performance of the TREND variable is largely unchanged. In equation 2 of Table A1.8 the model is re-estimated without the seasonal dummy variables. The coefficient on the spot crude

Table A1.5: The Relationship Between Production and Spot Crude Prices for the Brent Oilfield (January 1983–February 1985)

Dependent Variable	Constant	PRBREN	TREND	DW1	DSP1	DSM1	DW2	DSP2	DSM2	DW3	DSP3	DSM3	R^2	SEE	DWS
1. BRENT	−190.6 (395.6)	15.8 (11.5)	2.8 (1.7)	60.7* (27.9)	52.1** (27.5)	−94.7* (27.9)							0.59	47.8	2.25
2. BRENT	−22.3 (450.9)	12.1 (13.2)	2.2 (2.0)				35.9 (30.7)	−32.8 (33.8)	−109.8* (33.2)				0.43	56.4	1.53
3. BRENT	−305.7 (376.1)	21.2** (11.2)	3.0** (1.6)							24.9 (25.6)	−130.1* (28.9)	−88.9* (28.4)	0.62	45.8	1.61

Table A1.6: The Relationship Between Oil Production and Spot Crude Prices (July 1982–February 1985)

Dependent Variable	Constant	TREND	PRBREN	R^2	SEE	DWS
1. BRENT	317.6 (446.2)	2.1 (2.4)	0.02 (11.8)	0.00	75.4	0.57
2. FORTIES	627.1* (198.6)	−1.9** (1.1)	−3.7 (5.2)	0.07	33.6	2.05
3. REST	552.0** (308.1)	19.6* (1.7)	2.5 (8.1)	0.92	52.0	1.60

Note :   REST is defined as total UKCS oil production *less* production from the Brent and Forties fields.

Table A1.7: The Relationship Between Oil Production and Spot Product Prices (January 1980–December 1983)

Dependent Variable	Constant	TREND	Brent Netback	R^2	SEE	DWS
1. BRENT	29.2 (164.6)	7.7* (1.0)	1.7 (4.3)	0.77	56.9	0.62
2. FORTIES	312.1* (103.5)	−0.7 (0.6)	5.2** (2.7)	0.32	35.8	1.37
3. REST	862.6* (153.6)	13.5* (0.9)	0.3 (4.0)	0.93	53.1	1.08

Note :   The Brent netback refers to basic refining at Rotterdam

price variable becomes negative but remains highly insignificant. We try shortening the data period to see if there is a switch in behaviour towards the end of the period. These results are shown by equations 3–5 in Table A1.8. It is clear from these results that the production data for the total UKCS provide absolutely no evidence of crude price sensitivity. Finally, the spot crude price is replaced by the Brent netback variable (for basic refining at Rotterdam). As can be seen from equation 6, the netback variable possesses the expected positive coefficient but is again insignificant. The Brent netback for complex refining at Rotterdam is also tried without success and the result is not reported.

Table A1.8: The Relationship Between Total UKCS Production and Prices

	Data Period	Dependent Variable	Constant	PRBREN	Brent Netback	TREND	DW2	DSP2	DSM2	R^2	SEE	DWS
1.	Jul 1982–Feb 1985	NSO	1146.6* (497.3)	9.8 (13.0)		20.4* (2.5)	80.5* (38.6)	9.3 (42.6)	99.0* (37.3)	0.87	76.0	1.59
2.	Jul 1982–Feb 1985	NSO	1496.7* (579.3)	−1.2 (15.3)		19.8* (3.1)				0.78	97.9	1.10
3.	Jan 1983–Feb 1985	NSO	1728.8* (835.6)	−8.8 (24.4)		19.5* (3.6)				0.66	108.5	1.02
4.	Jun 1983–Feb 1985	NSO	1678.7 (1579.8)	−6.0 (41.6)		19.0* (8.0)				0.50	118.0	0.82
5.	Jan 1984–Feb 1985	NSO	2347.0 (2341.8)	−13.1 (56.7)		10.6 (14.5)				0.07	119.1	0.90
6.	Jan 1980–Dec 1983	NSO	1203.9* (221.9)		7.1 (5.8)	20.5* (1.3)				0.92	76.6	1.07

A1.4   UKCS Oil Production and Seasonal Demand

(a) UKCS Oilfields. It is possible that oil producers simply adjust output in response to exogenously determined seasonal variations in the demand for crude oil. In an attempt to capture seasonal demand variations we experiment with two series:

— UK gas consumption (UKGAS)
— UK refinery throughput (UKRT)

The data for these two series are obtained from *Energy Trends* for the period January 1980–December 1984.[2] Using these seasonal demand variables, various specifications of the basic model are tried and the results are presented in Table A1.9.

First, we regress Brent oil production on a constant, a linear trend variable and UK gas consumption (equation 1). The UKGAS variable is significant with the anticipated positive coefficient. However, in equation 2 the addition of the seasonal dummy variables leads to the insignificance of both the UKGAS variable and the seasonal dummies. The results for the Forties field show a similar pattern except that the seasonal dummy variables in equation 4 retain their significance (the Forties equation without the seasonal dummies is not reported here). As for the Rest of the UKCS, the UKGAS variable is insignificant with or without the inclusion of the dummy variables (see equation 6).

The equations are re-estimated with UK refinery throughput replacing UK gas consumption as the exogenous seasonal demand proxy. The coefficient on the throughput variable is significant for the Forties field and for the Rest of the UKCS (equations 5 and 7 in Table A1.9). However, the significance of some dummy variables would suggest that production is characterized by seasonal patterns that are independent of variations in oil demand (under the assumption that refinery throughput is an adequate proxy for oil demand). For the Brent field, the significance of the dummy variables in equation 3 confirms the seasonal behaviour of production. The coefficient on the UKRT variable is insignificant and, unexpectedly, negative. This is the case even when the seasonal dummy variables are omitted from the equation.

In conclusion, the performance of the throughput variable provides some tentative evidence that demand variations contribute to an explanation of seasonal production patterns for the UKCS oilfields with the exception of Brent. Nevertheless, the UKCS fields continue to exhibit seasonal behaviour independently of our proxy for crude oil demand. Of course, refinery throughput may not be an accurate measure of oil demand. Furthermore, there may be a problem of causality, i.e. higher production may lead to increased refinery throughput and not vice versa. More convincing evidence of a relationship between oil production and seasonal demand variations is provided by the performance of the UKGAS variable. The loss of its significance once the seasonal dummy variables are included in the equation would suggest that they are capturing the same effect.

---

[2] UK Department of Energy.

Table A1.9: The Relationship Between Production and Seasonal Demand (January 1980–December 1984)

Dependent Variable	Constant	TREND	UKGAS	UKRT	DW2	DSP2	DSM2	DW3	DSP3	DSM3	R^2	SEE	DWS
1. BRENT	55.1* (26.1)	6.2* (0.5)	17.2* (5.9)								0.76	61.2	0.55
2. BRENT	85.2 (55.6)	6.1* (0.5)	12.7 (12.4)						-33.5 (28.2)	-12.4 (36.0)	0.76	61.7	0.55
3. BRENT	315.0* (130.9)	5.9* (0.5)		-0.02 (0.02)				-5.9 (24.8)	-60.2* (23.6)	-53.0* (24.0)	0.76	61.3	0.63
4. FORTIES	521.1* (28.1)	-1.6* (0.2)	-2.0 (6.3)					21.4** (12.5)	-13.7 (14.2)	-37.4* (18.2)	0.55	31.2	1.89
5. FORTIES	340.1* (62.3)	-1.3* (0.2)		0.02* (0.01)				24.3* (10.9)	-1.7 (11.2)	-21.3** (11.4)	0.61	29.1	1.84
6. REST	324.4* (49.5)	15.5* (0.6)	4.1 (13.7)		12.6 (38.2)	43.6 (29.4)	19.3 (32.3)				0.93	75.9	0.44
7. REST	-98.1 (147.6)	16.2* (0.6)		0.06* (0.02)	2.2 (26.5)	46.0** (25.9)	33.7 (26.5)				0.94	70.3	0.74

Note : REST is defined as total UKCS oil production *less* production from the Brent, Forties, Ninian and Piper fields.

Hence, the seasonal dummy variables may reflect the effect on oil production of seasonal variations in demand.

(b) Total UKCS. We now test the hypothesis of a relationship between production and seasonal oil demand for the UKCS as a whole. The estimated equations are given in Table A1.10. With respect to the UKGAS variable, the results for the total UKCS correspond closely to those for the Brent field. The UKGAS variable is significant when related to oil production without the seasonal dummy variables (equation 1). It then loses significance once the seasonal dummies are included in the regression (see equation 2). As for the refinery throughput variable. it has a significant positive coefficient in equations 3 and 5. However, as with the Forties field, the seasonal dummy variables generally retain their significance, indicating that total UKCS production displays seasonal behaviour independently of demand variations.

A1.5   Temporal Production Patterns: Norway

(a) Total Norwegian Continental Shelf. The basic regression model relates oil production to a constant, a linear trend variable and a set of quarterly dummy variables. Equations 1–3 in Table A1.11 present the results on temporal production patterns for the total Norwegian Continental Shelf over the period January 1980–February 1985. In all three equations the significant coefficient on the TREND variable indicates a trend increase of 3400 b/d per month over the data period. The dummy variables provide only weak evidence of output seasonality. In equation 1 the winter and spring quarters are significant at the 10 per cent confidence level. Their coefficients indicate output increases of 59,500 and 58,100 b/d in the winter and spring quarters respectively, relative to the autumn quarter (August, September, October). The only other significant dummy (even at the 10 per cent confidence level) is DSM3 in equation 3. Its coefficient implies a 51,700 b/d decrease in output for the summer quarter relative to the autumn quarter. Overall, the results indicate a low output level in late summer/early autumn relative to the rest of the year.

(b) Norwegian Continental Shelf Oilfields. We now examine the temporal production patterns of the three major Norwegian Continental Shelf oilfields – Statfjord, Ekofisk and Eldfisk – as well as those for the Rest of the Norwegian Continental Shelf. The results for the Statfjord field are given by equation 4 of Table A1.11. The TREND variable is significant and its positive coefficient shows a production increase of 5900 b/d per month. There is no evidence of seasonality in production from the Statfjord field. The dummy variables are always highly insignificant.

The results for the Ekofisk field (equation 5 of Table A1.11) indicate a slow downward trend in production over the data period. The significant coefficient on the TREND variable indicates a fall of 1100 b/d per month. The performance of the DW1 and DSP1 variables reveals some weak seasonality. Both coefficients are significant at the 10 per cent confidence level. Nevertheless, they imply rises in production of 13,200 and 15,300 b/d in the winter and spring quarters

Table A1.10: The Relationship Between Total UKCS Oil Production and Seasonal Demand (January 1980–December 1984)

	Constant	TREND	UKGAS	UKRT	DW1	DSP1	DSM1	DW2	DSP2	DSM2	DW3	DSP3	DSM3	R^2	SEE	DWS
1.	1357.3* (31.7)	19.3* (0.6)	32.9* (7.2)											0.95	74.2	1.30
2.	1405.8* (67.9)	19.4* (0.6)	20.2 (15.2)								22.4 (30.3)	−21.8 (34.5)	−28.4 (44.0)	0.95	75.4	1.32
3.	1085.3* (157.8)	20.1* (0.6)		0.05* (0.02)	52.1** (29.2)	94.3* (27.1)	5.7 (27.1)							0.96	73.7	1.25
4.	1282.1* (157.6)	19.8* (0.6)		0.03 (0.02)				51.7** (28.3)	24.8 (27.6)	−53.6** (28.3)				0.95	75.1	1.22
5.	1214.6* (159.4)	19.9* (0.6)		0.04** (0.02)							44.6 (28.0)	−34.0 (28.7)	−55.4** (29.3)	0.95	74.6	1.36

Note :   The dependent variable is NSO in all cases.

Table A1.11: Temporal Production Patterns for the Norwegian Continental Shelf (January 1980–February 1985)

	Dependent Variable	Constant	TREND	DW1	DSP1	DSM1	DW2	DSP2	DSM2	DW3	DSP3	DSM3	R^2	SEE	DWS
1.	NNS	449.0* (29.8)	3.4* (0.6)	59.5** (30.6)	58.1** (30.7)	26.3 (31.1)							0.35	85.1	0.98
2.	NNS	470.7* (30.3)	3.4* (0.6)				44.7 (30.3)	27.5 (31.4)	−14.6 (31.3)				0.34	85.5	0.99
3.	NNS	491.5* (30.4)	3.4* (0.6)							17.0 (30.1)	5.7 (31.1)	−51.7** (30.9)	0.36	84.6	0.98
4.	STAT	26.0 (26.0)	5.9* (0.5)	24.0 (26.8)	20.3 (26.8)	14.9 (27.2)							0.66	74.4	0.51
5.	EKO	195.8* (7.6)	−1.1* (0.2)	13.2** (7.8)	15.3** (7.8)	1.0 (7.9)							0.48	21.7	1.65
6.	ELD	101.8* (4.7)	−0.8* (0.1)	4.4 (4.8)	4.9 (4.8)	2.8 (4.9)							0.54	13.4	1.87

respectively, relative to the autumn quarter. The results on seasonality for the Ekofisk field correspond closely to those for the Norwegian Continental Shelf as a whole.

For the Eldfisk field, a slight downward trend is shown by the significant negative coefficient on the TREND variable (equation 6 of Table A1.11), indicating a drop of 800 b/d per month over the data period. However, Eldfisk production gives no evidence of seasonality. All the dummy variables are insignificant.

The results on seasonal production patterns for the Rest of the Norwegian Continental Shelf are presented separately in Table A1.12. Production for the Rest of the Norwegian Continental Shelf is simply obtained by taking the difference between total Norwegian Continental Shelf oil production and the combined production of the three major Norwegian oilfields – Statfjord, Ekofisk and Eldfisk.

Equation 1 of Table A1.12 provides estimates for the whole data period. The significant coefficient on the TREND variable shows a slight downward trend in output of 500 b/d per month on average. In addition, there is some weak evidence of output seasonality. The coefficients on DW1 and DSP1 are significant at the 10 per cent confidence level and indicate that there are output increases of 17,800 and 17,700 b/d in the winter and spring quarters respectively, relative to the autumn quarter.

The model is estimated separately for two sub-periods, January 1980–June 1982 and July 1982–February 1985, because a plot of production shows that these periods are characterized by contrasting trends. For the first period (equation 2), the TREND variable captures a strong downward trend in output of 3200 b/d per month. In this period, the strongest performer among the dummy variables is DSM3 which is significant at the 5 per cent confidence level. Its coefficent implies an output fall of 24,500 b/d in the summer quarter (July, August, September) relative to the autumn quarter. The result reflects large production falls in July 1980 and August 1981. In contrast, the results for the second period (equation 3) show an upward trend in output. The coefficient on the TREND variable is significant and represents average output growth of 1000 b/d per month over the period. Once again, the seasonal dummies perform more strongly over the shorter period. The significant coefficients on the DW1 and DSP1 variables imply output rises of 16,100 and 19,800 b/d in the winter and spring quarters respectively, relative to the autumn quarter (August, September, October).

A1.6 Norwegian Continental Shelf Oil Production and Prices

To test whether the seasonal production patterns for the Norwegian Continental Shelf are explained by economic factors we estimate a model in which production is related to a constant, the TREND variable, the seasonal dummy variables and the spot crude price. As a general indicator of Norwegian oil prices we use the Ekofisk monthly average spot price. This series is obtained from *International Crude Oil and Product Prices* for the period January 1980–January 1985.[3] Under the assumption that price is exogenous, the model is estimated for each individual field and for the

[3] Middle East Petroleum and Economic Publications.

Table A1.12: Temporal Production Patterns for the Rest of Norwegian Continental Shelf

Data Period	Dependent Variable	Constant	TREND	DW1	DSP1	DSM1	DW3	DSP3	DSM3	R^2	SEE	DWS
1. Jan 1980–Feb 1985	NREST	125.3* (9.1)	−0.5* (0.2)	17.8** (9.3)	17.7** (9.3)	7.5 (9.5)				0.12	25.9	0.87
2. Jan 1980–Jun 1982	NREST	180.7* (10.9)	−3.2* (0.4)				2.1 (10.6)	1.6 (10.5)	−24.5* (11.6)	0.67	19.9	2.61
3. Jul 1982–Feb 1985	NREST	58.4* (11.9)	1.0* (0.2)	16.1* (6.1)	19.8* (6.5)	9.9 (6.4)				0.46	12.8	1.02

Note :   NREST is defined as total Norwegian Continental Shelf oil production *less* production from the Statfjord, Ekofisk and Eldfisk fields.

Table A1.13: The Relationship Between Norwegian Continental Shelf Oil Production and Prices

Data Period	Dependent Variable	Constant	TREND	Spot Crude Price	Netback	R^2	SEE	DWS
1. Jan 1980–Jan 1985	NNS	307.6 (246.4)	4.3* (1.4)	4.6 (6.2)		0.30	87.6	0.92
2. Jan 1980–Jan 1985	EKO	155.4* (63.3)	−0.9* (0.4)	1.2 (1.6)		0.44	22.5	1.54
3. Jan 1980–Jan 1985	NREST	152.2 (223.0)	5.2* (1.3)	3.3 (5.6)		0.49	79.3	0.71
4. Jan 1980–Dec 1983	NNS	398.1 (250.5)	1.7 (1.5)		3.5 (6.6)	0.03	86.5	1.11
5. Jan 1980–Dec 1983	EKO	230.7* (70.2)	−1.5* (0.4)		−0.6 (1.8)	0.38	24.3	1.52
6. Jan 1980–Dec 1983	NREST	167.5 (226.2)	3.2* (1.4)		4.1 (5.9)	0.14	78.1	0.78

Note : NREST is defined as total Norwegian Continental Shelf oil production *less* production from the Ekofisk field.

Norwegian Continental Shelf as a whole. The results are not reported because the inclusion of the spot crude price provides no additional explanatory power to the model. Nor has it any effect on the performance of the other variables. We now drop the seasonal dummies, therefore, to see if there is any improvement in the performance of the spot crude price.

The results for the total Norwegian Continental Shelf and the Ekofisk field are given by equations 1 and 2 respectively, in Table A1.13. The spot crude price is again insignificant in both cases. A similar result is obtained for the Rest of the Norwegian Continental Shelf which now includes Statfjord and Eldfisk production (see equation 3).

Finally, we test to see whether there is a relationship between production and spot product prices. However, we have no series to act as a general indicator of Norwegian netback values. We are obliged, therefore, to use the Brent netback series provided by an industry source for the period January 1980–December 1983. This is probably not too serious a problem as we would expect movements in all North Sea netbacks to be very closely correlated. In any case, equations 4–6 of Table A1.13 show that for the Norwegian Continental Shelf there is no significant relationship between crude oil production and netbacks (for basic refining at Rotterdam) over the data period. We also tried the netback for complex refining at Rotterdam but without success and the results are not reported. The failure to explain seasonal production patterns with price variables is perhaps not too surprising given the weakness of the seasonality.

ANNEX 2

THE GAINS FROM TAX SPINNING

The formal proof that an overall fiscal advantage obtains whenever an integrated company puts the market between its North Sea supplies and the demand of its refineries, so long as the arm's length (say spot) price is lower than the BNOC price (or the price at which the Inland Revenue is likely to assess non-arm's length sales) is as follows.

Let t_p and t_c be the rates of PRT and Corporation Tax respectively;
p_o and p_s the BNOC and spot prices of crude oil respectively;
p_f the price of a composite barrel of refined products;
c_u the per barrel cost of production of crude oil upstream;
c_d the per barrel cost downstream;
c_s the transaction cost involved in selling and then buying a barrel of crude oil on the spot market.

OPTION I is the appropriation by an integrated oil company of its UKCS crude for use in its own refineries or the sales of this oil to affiliates;

OPTION II is 'spinning', that is, arm's length sales by the integrated oil company of its UKCS crude and purchases of North Sea and/or other crudes for own refinery or affiliates' requirements.

Under OPTION I the tax liability (per barrel) other than royalties is:

$$(p_o-c_u)t_p + [(p_o-c_u)-(p_o-c_u)t_p]\ t_c + (p_f-p_o-c_d)t_c \tag{2.1}$$

This reduces to:

$$(p_o-c_u)(t_p+t_c-t_pt_c) + (p_f-p_o-c_d)t_c \tag{2.2}$$

Because of the ring-fence, a downstream loss cannot be set against an upstream profit. Thus, the term $(p_f-p_c-p_d)t_c$ is only relevant when positive. This term drops out of the equation when there are downstream losses, and in this special case the tax liability under OPTION I becomes:

$$(p_o-c_u)(t_p+t_c-t_pt_c) \tag{2.3}$$

For simplicity, we have assumed that there are no allowances against PRT and Corporation Tax other than the costs c_u and c_d. Introducing these allowances would complicate the equations but does not alter the main result.

The tax liability (per barrel) of an integrated company that takes OPTION II, that is, spins its oil, is:

$$(p_s - c_u)(t_p + t_c - t_p t_c) + (p_f - p_s - c_d)t_c \qquad (2.4)$$

or:

$$(p_s - c_u)(t_p + t_c - t_p t_c) \qquad (2.5)$$

when there are downstream losses.

The difference in tax liability between OPTION I and OPTION II, in general, is $(2.2) - (2.4)$. Subtracting, we obtain:

$$(p_o - p_s)(t_p + t_c - t_p t_c) - (p_o - p_s)t_c \qquad (2.6)$$

Rearranging (2.6) gives:

$$(p_o - p_s)(1 - t_c)t_p \qquad (2.7)$$

It can be immediately seen that (2.7) is a positive number when $p_o - p_s$ is positive, that is, if the spot price is lower than the BNOC price. This is because $1 - t_c$ is always positive, given that marginal taxation rates are less than one. In other words, direct appropriation of oil or non-arm's length sales involve a greater tax liability than spinning whenever the spot price is below the BNOC price.

Note that the existence of a tax gain in spinning oil depends solely on the condition $p_s < p_o$. It is often said that spinning involves a tax advantage when $p_s < p_o$ because the PRT rate is higher than the Corporation Tax rate. This is a loose statement. A gain would obtain whether the corporation rate was lower or higher than the PRT rate, as is obvious from equation (2.7). What matters is that the overall taxation rate is higher upstream than downstream.

For any given $p_o - p_s > 0$, the tax gain is greater the lower is t_c and the higher is t_p. For any given $p_o - p_s > 0$ and a given t_p, the tax gain is the highest when $t_c = 0$. The other extreme case is that of zero gain when $t_p = 0$.

Thus, differences between the tax rates t_p and t_c may alter the strength of the incentive to spin, but save in the extreme cases they do not prejudice the existence of tax gains.

Consider now the case when the oil company pays no downstream corporation taxes because of downstream (tax) losses. In this situation, the difference between the tax liability incurred under OPTIONS I and II is given by $(2.3) - (2.5)$. Subtracting, we get:

$$(p_o - p_s)(t_p + t_c - t_p t_c) \qquad (2.8)$$

or:

$$(p_o - p_s)(1 - t_c)t_p + (p_o - p_s)t_c \qquad (2.9)$$

This, however, assumes that both OPTIONS I and II involve downstream tax losses. A case may arise in which there is no corporation tax liability downstream under OPTION I but liabilities are incurred under OPTION II. The reason is that the refineries would buy their crude oil inputs under OPTION II at a price p_s lower than the transfer price p_o at which they would buy under OPTION I. In such a case, the difference in tax liability between OPTIONS I and II is:

$$(p_o - p_s)(t_p + t_c - t_p t_c) - (p_f - p_s - c_d)t_c \tag{2.10}$$

or:

$$(p_o - p_s)(1 - t_c)t_p + (p_o - p_s)t_c - (p_f - p_c - c_d)t_c \tag{2.11}$$

Compare now the situation in which there is downstream tax liability (when the tax gain from spinning is given by (2.6) or (2.7)) with the situation in which there is no downstream liability at all (formulae (2.8) and (2.9)) and with the situation in which OPTION I carries no downstream tax liability but OPTION II does (formulae (2.10) and (2.11)). It is clear that (2.9) is greater than (2.7) by an amount equal to $(p_o - p_s)t_c$. There is, therefore, a greater tax advantage to be obtained from the spinning under both options when no corporation tax is paid downstream than when there is a downstream tax liability.

But it is not immediately apparent that the difference between tax liabilities under OPTIONS I and II when there is no downstream tax liability under OPTION I, (2.10), is greater than the difference in tax liability in the general case, (2.6). For this to be so, the following condition must be satisfied:

$$(p_f - p_s - c_d)t_c < (p_o - p_s)t_c \quad \text{or} \quad p_f - c_d < p_o \tag{2.12}$$

We know, however, that no downstream corporation tax liability arises in this case under OPTION I when crude oil inputs are charged at price p_o to the refineries. This means that we already have:

$$p_f \leqslant p_o + c_d \quad \text{or} \quad p_f - c_d \leqslant p_o \tag{2.13}$$

But this is the same as (2.12). The conclusion is that not only does it pay to spin (given $p_s < p_o$) when spinning creates a downstream corporation tax liability that does not exist under OPTION I, but also the tax gain is then greater than in the general case.

The difference in tax liability between OPTIONS I and II does not necessarily make spinning profitable if spinning involves certain costs. Transaction costs are indeed incurred since the oil company engages in two market operations: selling the oil it produces and buying its own requirements. For an overall gain to obtain, the tax gain must be higher than the transaction costs (net of tax). In other words:

$$(p_o - p_s)(1 - t_c)t_p > c_s(1 - t_c)$$

or $\quad (p_o - p_s) > \dfrac{c_s}{t_p}$ $\tag{2.15}$

Assuming $c_s = 15$ cents, $t_p = 0.75$, we obtain:

$$p_o - p_s > 20 \text{ cents} \tag{2.16}$$

This is a very remarkable result which indicates that a very small difference between spot and BNOC prices is sufficient to create an overall gain from spinning.

The proof that spinning always involves a tax advantage when $p_s < p_o$ is established on the assumption that PRT on retained oil is calculated on the basis of the BNOC term price p_o. We suggested that at some time in 1984 the Inland Revenue took the view that p_o was no longer representative of the market price. The question is whether the Revenue would ever accept p_s as the representative market price, or would always argue in favour of a 'tax-relevant' market price p_m lying somewhere between p_s and another (higher) price.

We have shown that the only necessary condition for a pure tax gain to arise from spinning as against direct appropriation of oil by integrated oil companies is:

$$p_s < p_o \tag{2.17}$$

and that the only necessary condition for an overall pure gain to arise is:

$$p_o - p_s > \frac{c_s}{t_p} \tag{2.18}$$

It follows that:

— Spinning by integrated companies may cease to be preferable to direct appropriation should the OTO consider the spot price of the day, p_s, to be the relevant price for internal transactions.
— However, spinning would still remain profitable if the OTO opted for a price $p_m > p_s$, provided that:

$$p_m - p_s > \frac{c_s}{t_p} \tag{2.19}$$

i.e. so long as the tax market-equivalent price, say, is more than 20 cents above the spot price.

ANNEX 3

A MODEL OF THE EQUILIBRIUM NUMBER OF FORWARD DEALS IN A MARKET WITH HEDGERS AND SPECULATORS

A3.1 Introduction

This model attempts to find a simple way of analysing the determinants of the number of deals in forward transactions in a market characterized by the presence not only of 'physical' sellers and buyers of the product but also of 'paper' sellers and buyers. In particular, attention is given to the case where the number of paper deals is greater than the number of physical units available. Such a situation is explained purely by speculative motives of *both* parties to the extra deals. An important dimension of the analysis is the introduction of a distribution of beliefs about future prices: unless we can find two agents with sufficiently different beliefs there can be no purely speculative deals because this is the only reason for exchange (there is no comparative advantage in production between them).

We begin with a characterization of the agents and their behaviour, then move to an analysis of the equilibrium and a discussion of its implications within the model, and finally we relate the insights gained to the Brent oil market.

A3.2 The Agents

There are three classes of agents in the model:

— physical producers who are selling the commodity in question
— physical buyers ('consumers') who are buying the commodity as an input
— traders who buy and sell the commodity or paper claims to it if they expect to make sufficient profit by so doing

We discuss each agent in turn (the first and second classes are symmetric so that most of the relevant discussion will be given for the former).

(a) Physical Producers. The producers are characterized by a constant stream of production which we take to be certain and, in the short run, completely inflexible (i.e. it is too expensive to make short-run adjustments within a month or so if the price appears likely to be unfavourable in the near future). Hence, given fluctuations in demand (and perhaps unforeseen fluctuations in supply), the producer faces a world of price uncertainty on a fixed quantity.

In such cases most commodity markets have tended to evolve some system of reducing the uncertainty, although other parties will only agree if there is some compensation for them. A long-term contract system between physical buyers and sellers (both facing short-term price variability) is one way of achieving this, but the very inflexibility of such contracts has a very large cost if the 'equilibrium' price shifts. Hence there is a desire for some certainty but with an element of flexibility.

The emergence of a forward market or a futures market is an intermediate position which is also very common in commodity trading.[1] The essential feature of the forward market is that the seller can choose today (this month) between *waiting* to sell tomorrow's production on the spot market when it is produced, at whatever is then the ruling price, or selling it forward *now* at a price agreed today. If he sells a physical 'lot' for a certain forward price he has 'hedged' against the spot price uncertainty.

The critical aspect of any such deal is that somebody else must be willing to buy forward at that price. This could either be a genuine user of the commodity, who has a continuing need for it, or a 'middleman', who has no need for the commodity as such. Since a user of the commodity also faces uncertainty, he may be willing to enter a 'buy-hedge' to reduce uncertainty. In such a case both parties may find a 'forward' price that is acceptable to both and reduces their perceived risk. In practice, this system of a series of constantly changing short-run contracts with no middlemen does not seem to be widespread. This is presumably because the constant shopping around for possible partners is costly and requires specialist knowledge. The producers and 'consumers' could of course evolve such specialization and some no doubt do so, but if there are substantial entry costs and then economies of scale, they may feel that it is not worth trying to extend their comparative advantage to this activity.

It is this that allows the deals with middlemen to emerge. A trader would need to buy forward from a producer and sell forward separately to a 'consumer'. Since these cannot be done simultaneously there is clearly a risk involved, so the trader will only undertake the risk if the margin between the deals is adequate compensation. Hence we see the coincidence of interests arising.

The producer will be prepared to sell-hedge at a price below the spot price expected by him as a premium for the certainty gained – he sells to a trader who may, for example, expect the same spot price and will hence buy-hedge at the lower price and take on the uncertainty together with the expected profit that will accrue. Of course the trader, if he has the same expectation of the spot price as the producer, will only buy if his actual risk is lower or if he is less risk averse (more prepared to take a risk on the same sum of money) than the producer. We can also include the costs of searching for an alternative in the risk premium – if these are lower for the trader (as we argued by pointing to his comparative advantage arising from specialization), then this provides a further reason for the one to buy and the other to sell at the same forward price and at the same expected spot price for the next period.

To sum up; we see that if a producer i expects 'on average' that the spot price will be P_i^e then he will be willing to hedge at any forward price above $P_i^e - R$, where R is the risk premium (assumed to be equal for all producers).

---

[1] In a forward market deals are done with other traders at a mutually agreed price while in a futures market deals are done with a central agency that sets all prices.

(b) Physical Consumers. The case of consumers we take to be largely symmetrical to that of producers. They face the need to buy steadily with little ability or desire to vary the quantities they use in the short run if the spot price varies. Hence they face a constant demand for inputs which vary in price on the spot market over time. A forward purchase can reduce the uncertainty so there is a motive to 'buy-hedge' at a forward price *above* the expected (average) spot price for the next period. The extra amount a consumer is willing to pay is determined by his risk premium R, and for simplicity we assume this to be the same as that of the producers.

Again the high search costs and the high risk premium make it possible for a consumer to be willing to buy-hedge from a trader even when both expect the same average spot price.

(c) Traders. As we have argued, traders are a class of agents who have no use for the commodity as such but who, because of their ability in searching for deals and their lower risk aversion, can intertemporally bring buyers and sellers of a commodity together through the indirect paper transaction of hedging.

In the simplest case, where agents of every type realize that prices are uncertain but perceive the same expected (mean) value, there is still a role for the traders if their costs and risk premiums are lower than the primary agents'. A very important insight to be gained from this extreme case is that *on average* the primary agents are transferring resources to the traders in return for the services rendered. It can be seen that once there is a motive to deal forward it is inevitable (and beneficial) for traders, who have lower costs, to enter the market.

In a certain sense, even in this very specialized case, the traders are 'speculating' since they must inevitably have an 'open' position for at least a short period of time while they attempt to match a sell-forward deal to the buy-forward deal already acquired (or vice versa). To take such a position is to speculate on the price at which the second half of the deal can be made, i.e. until the deals are matched the position has become more risky than that held before the open position was taken. (The open position can be seen as the trader's counterpart to a hedge for the physical owner of a commodity.) Hence, hedging requires speculators and indeed pays them if their costs and risk premiums are low enough.

However, there is a second aspect to speculation which is in practice indistinguishable from the first. Two agents on a given day may well expect different mean prices to rule in the future – if so there is a chance for a deal between them. The one expecting a high spot price in the future will be prepared to buy forward at a lower than expected spot price while the one expecting a low spot price will be prepared to sell forward at a price higher than the spot price he expects. Thus there may be an intermediate forward price at which both parties expect to make a profit. Of course such deals are risky and each must expect to make at least his risk premium on the deal to make it worthwhile. Hence, if expectations are more than twice the risk premium apart, there exists a forward price at which both will willingly enter a deal. The crucial aspect of such a deal is that *ex post*, once the actual spot price has emerged, the total profit to the two agents will be zero. One will gain and the other will lose. Pure speculation of this type merely redistributes resources among speculators.

The result of analysing these activities separately is that we can now see that the total resources to be divided among traders are provided solely by the hedging of producers and consumers – however this is shared over all speculative activity – whether merely that of taking on an increased risk position (because of differential costs or attitudes to risk) or also that arising from different expectations of the immediate future spot price. Thus the amount of hedging will determine the total amount of trading in the forward market by limiting the average profit per deal available to the market as a whole. We can imagine the normal entry/exit procedure of a competitive market applies to trading. Over the long run, the average margin per deal must equal the risk premium for taking on the deal. If any firm persistently outguesses its rivals it will make extra profits but this must entail others making the corresponding losses. Such a situation cannot be sustained – loss makers will exit from the market until the average is restored. A persistently accurate player will eventually find fewer players prepared to deal with him, and a process of adjustment will take place.

We see then that traders will take on the risks from hedgers and will also be willing to speculate. However the *long-run* gains from speculation are obtained solely from the margin paid by the hedgers for security.

An obvious feature of the traders' strategy will be the attempt to close their positions as rapidly as they can, consistent with the expectation of making a profit. We can expect their comparative advantage in searching to result in deals being very frequent. With a large number of players and a lengthy forward period there is the opportunity for a long chain of links to be established before the physical delivery takes place.

A3.3   Long-run Equilibrium for Wet and Paper Deals

We characterize the market by a number of assumptions which are not likely to hold over a lengthy period of time but which may hold over the medium term. Our analysis concentrates on such a market once agents have learnt their roles and adjusted to the market.

(a)   There are a number of producers who have a total quantity Q to sell either spot or forward. Each producer has one unit to sell per period, the size of which is small relative to the total quantity.

(b)   There are a number of refiners who have a total quantity Q to buy. Again purchases are in units of small size relative to the total, and each refiner must buy one unit per period.

(c)   Spot prices are random, but over the long run they have a mean P^*.

(d)   Producers and refiners are risk averse and are willing to pay a risk premium of R per unit to achieve certainty on the forward market. We assume this premium to be invariant over the actual range of spot prices experienced and over variations of other factors in the market.

(e)   At a given moment of time t, each producer i has an expectation of the spot price for the next period P_i^e. Hence producer i is willing to sell at a forward price at or above:

$$P_i^e - R \tag{3.1}$$

The price expected in any period is independent of that expected in other periods (i.e. nobody is consistently an optimist or a pessimist).

(f) The distribution of expectations in any period is described by a rectangular distribution over the range:

$$P^* - D \text{ to } P^* + D \tag{3.2}$$

The total range of the distribution ($2D$) is a parameter.

(g) Refiners are willing to buy forward at prices at or below:

$$P_j^e + R \tag{3.3}$$

where P_j^e is the price expected by the jth refiner.

(h) The distribution of refiners' expectations is also rectangular with range $2D$ and mean P^*.

(i) The number of traders in the market is n. Traders have a risk premium of r ($r<R$) for buying or selling forward and their expectations of future spot prices are characterized by a rectangular distribution with range $2d$ centred at P^*. A trader is only large enough to buy or sell at most one unit per period.

This simple characterization of the market needs to be carefully interpreted. Each group of agents as a whole has the correct average expectation of the future spot price. Nevertheless, there are always some expecting higher and some expecting lower prices – these individuals change their views randomly with respect to the mean price. If the expectations of two agents are sufficiently far apart then they are willing to do a deal that covers the difference in risk premiums.

(a) Supply. The supply of 'sell-hedges' is determined by the actions of two groups: first the producers and secondly certain traders who think that at the ruling forward price they will be able to make a profit.

The supply of 'wet hedging' at any forward price P^f is thus the number of producers who are willing to hedge at that price, i.e.:

$$Q \int_{P^*-D}^{P^f+R} f(P^e)dP^e \tag{3.4}$$

where $f(P^e)$ is the proportion of producers expecting a future spot price of P^e. This supply is the total number of producers who expect future spot prices to be below the forward price plus risk premium. (The lower limit of the integral is the minimum price expected by any supplier.) For example, if the forward price is 10 and the risk premium is 2 then a producer will hold in order to sell spot only if he a believes the future spot will be 12 or more.

Given that we have assumed a rectangular distribution of range $2D$, the integral gives the total supply of wet hedging as:

$$\frac{Q(P^f+R-P^*+D)}{2D} \tag{3.5}$$

For paper deals the argument is not exactly the same. The seller of a paper hedge is not considering the *selling* of a wet (future) contract as the risk factor to be considered but rather the *buying* of a future contract to close the position as the uncertain part of the deal. A trader will sell forward if the price he expects, P^e, is less than the forward price by at least the risk premium r. Hence the total supply of paper sales is:

$$n \int_{P*-d}^{P^f-r} f(P^e)dP^e \tag{3.6}$$

$$= \frac{n(P^f-r-P*+d)}{2d} \tag{3.7}$$

(b) Demand. The demand for 'wet hedges' can be analysed by similar techniques. Refiners will be willing to buy forward if they expect the future spot price to be above the forward price less the risk premium. For example, at a forward price of 10 and a risk premium of 2 they will be prepared to hedge if they expect the spot price to be greater than 8. Hence the demand for wet hedging is:

$$Q \int_{P^f-R}^{P*+D} f(P^e)dP^e \tag{3.8}$$

$$= \frac{Q(P*+R+D-P^f)}{2D} \tag{3.9}$$

Finally, by arguments modified in a similar way to those for the supply of paper hedging, we have the total demand for paper deals:

$$\frac{n(P*-r+d-P^f)}{2d} \tag{3.10}$$

This simple analysis leads to linear supply and demand schedules which depend on the different expectations of the agents. Each function starts at zero and finishes at Q (or n). Market clearing is achieved by a single forward price. We see that the opportunity for a deal at a given forward price is created because the difference in expectations is wider than the risk premium. For example, with no risk premium half the agents would sell to the other half because only at the 'median' price would the number expecting the future price to be higher be balanced by the number of those expecting it to be lower so that all could be satisfied. As the risk premium rises, expectations must differ from the favoured price in order to balance supply and demand. Figure A3.1 illustrates the model for 'wet deals' only.

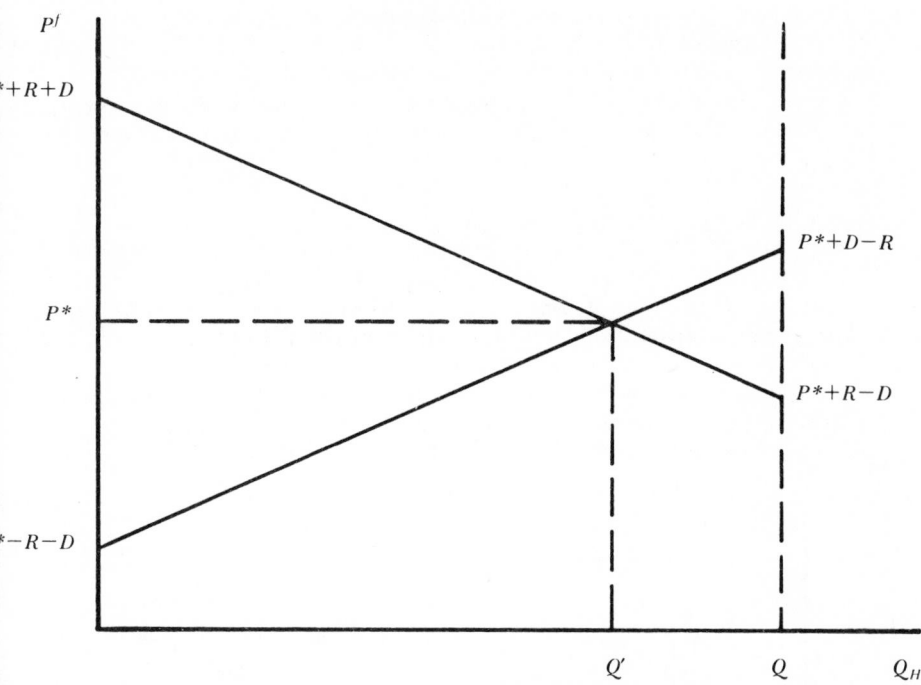

Figure A3.1 Supply and Demand for Wet Deals

The minima of the demand and supply are easily found by solving for Q_H (quantity hedged) equal to zero. The upper limit (all hedged) of Q is found by equating Q_H to Q. The first feature to notice is that the curves will intersect only if:

$$D - R > R - D$$

i.e. $D > R$ (3.11)

If $D < R$ then demand always exceeds supply and everything is hedged.

If $D > R$ then only a fraction of total output is hedged. The price that clears the market is of course P^* (the long-term average spot price), since, with symmetry of beliefs above and below this price and on both sides of the market, the number of deals will be equated at this price. At such a price the marginal seller expects the future price to be $P^* + R$ while the marginal buyer expects it to be $P^* - R$. The equilibrium quantity hedged is given by noting that at price $P^* - R - D$ nothing is supplied and at $P^* + D - R$ quantity Q is supplied, i.e. $Q/2D$ supply per unit price rise. Since the equilibrium price rise is $R + D$ the equilibrium hedged quantity is:

$$Q' = \frac{Q(R+D)}{2D} \qquad\qquad (3.12)$$

The lower the risk premium is in relation to the spread of values, the less is traded forward. In a riskless market with an equal distribution of views, 50 per cent will be hedged – the buyers expecting a higher than average price will do a forward deal with the sellers expecting a lower than average future price.

A more difficult question in analysing a market when there are only primary buyers and sellers is that of the transfer of resources. We imagine that every lot is bought and sold just once so that the 'profits' on hedging are defined solely in terms of opportunity costs. Once we move to a market with pure traders this changes radically.

A second difficulty is describing the set of prices at which trade takes place. If there were perfect information, then P^f would also be the only price at which trade took place. However, in a forward market (as opposed to a futures market), information is not perfect and transactions may well take place at other prices. This avenue will also be explored once we add traders to the picture.

If we add in the supply and demand from the traders (where gains come from the speculative element created by different expectations and from taking on hedges because of having a lower risk premium) we obtain the supply and demand schedules for 'paper' deals.

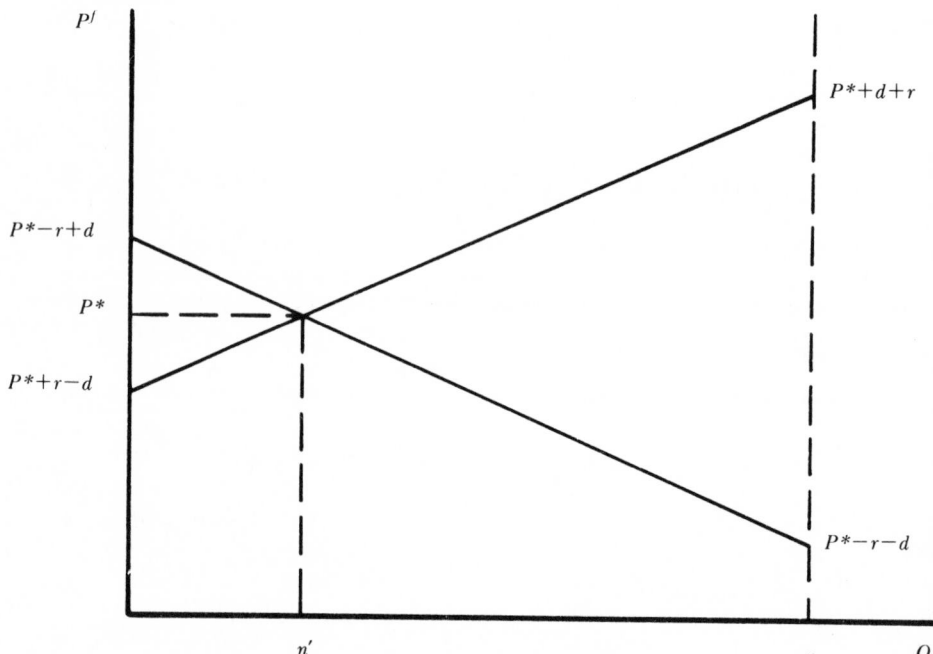

Figure A3.2   Supply and Demand for Paper Deals

The schedules are exactly analogous to those for wet deals (see Figure A3.2). The maximum trade would be reached at n – where all firms in the market are actually making a deal (in both directions). Under our special assumptions about the mean expected future price we see that a price of $P^f = P^*$ will clear the market in any period and that the number of deals done is:

$$n' = n\frac{(d-r)}{2d} \; ; \; r \leq d \tag{3.13}$$

Provided that the risk premium is less than half the range of expectations, the number of deals is less than n, i.e. not all firms who exist in the market will do a deal in each period. However, over the long run, their expectations will shift and they will do (n'/n) deals per period.

Because they clear at the same marginal price, the two markets can now easily be put back together and we can see that the average number of links in the daisy chain C is:

$$C = \frac{n'}{Q'} + 1 \tag{3.14}$$

or:

$$C = \frac{\dfrac{n(d-r)}{2d} + \dfrac{Q(R+D)}{2D}}{\dfrac{Q(R+D)}{2D}} \tag{3.15}$$

This leaves open the critical question of the determination of n. This has an important effect in our model because with expectations spread evenly over the number of traders, the greater is n the more paper deals will be possible (pairs of optimists and pessimists). We can attempt a crude model of the determination of n in equilibrium by noting that on balance the speculative activities in the market contribute no net resources – only the hedging does this. Hence the total resources to be shared among all speculative deals is the total paid by the hedgers. This must be enough to support the traders – i.e. on average each trader makes nothing from speculation but obtains a share of the resources provided by the primary users. This must be enough on average to support the number of speculative deals made by the trader.

If there were perfect discrimination by the traders against primary agents – each selling the hedge for the lowest price at which he was willing to enter the market or buying a hedge at the highest price – then the maximum transfer would be the integral between the supply and demand curves of the wet traders (up to the market clearing price).

Hence the total resource π is the number of wet deals times average margin between supply and demand:

$$\pi = Q'(R+D) \tag{3.16}$$

Hence:

$$\pi = \frac{Q(R+D)^2}{2D} \tag{3.17}$$

The margin per paper deal to each party of the deal is therefore:

$$M = \frac{\pi}{n'} \tag{3.18}$$

$$= \frac{Q(R+D)^2}{2D} \bigg/ \frac{2n(d-r)}{2d} \tag{3.19}$$

For an equilibrium to exist the number of firms n must be such that the average per deal made is the risk premium r.

Therefore:

$$r = \frac{Q(R+D)^2}{2D} \bigg/ \frac{2n(d-r)}{2d} \tag{3.20}$$

gives the equilibrium value for n (under perfect discrimination), i.e.:

$$n = \frac{Q(R+D)^2 d}{2Dr(d-r)} \tag{3.21}$$

Some features of this are self evident – by simple inspection we see that n increases as:

— the total wet availability increases;
— the divergence of views D of primary agents increases;
— the risk premium of primary agents increases.

The effect of the divergence of views of traders cannot be predicted *a priori*.

Some special cases are also of note. If the spread of expectations is the same for both classes of agent ($d = D$), then:

$$n = \frac{Q(R+D)^2}{2r(D-r)} \tag{3.22}$$

From (3.21) we see that the equilibrium length of the daisy chain per period of time is in general:

$$C = \frac{\dfrac{Q(R+D)^2 d}{2Dr(d-r)} \cdot \dfrac{d-r}{2d} + \dfrac{Q(R+D)}{2D}}{\dfrac{Q(R+D)}{2D}} \tag{3.23}$$

$$= \frac{R+D}{2r} + 1 \tag{3.24}$$

so that the length of the average daisy chain is invariant in equilibrium to the quantity of base resource, while the percentage of traders active at any time (n'/n) is:

$$\frac{d-r}{2d} \tag{3.25}$$

These results are all contingent on the assumption of perfect discrimination against primary agents. As this part of the market becomes better informed, the degree of discrimination should steadily decrease and with it the net transfer of resources. In the limit all secondary agents would be squeezed out if the marginal price for a forward contract were also the average price.

Finally, the number of paper deals in the market is in general:

$$n' = \frac{Q(R+D)^2}{4Dr} \tag{3.26}$$

A3.4 Comments and Interpretation

This is the simplest type of model with two types of agents. The primary buyers and sellers have the same range of expectations and the same risk premium – the traders have a different range and a different risk premium. By assuming:

(a) equilibrium so that on average every deal makes a sufficient profit for the traders;

(b) the expected price (averaged over all agents) is constant and equal for the two classes of agents;

(c) the distributions of expectations are characterized by stable rectangular distributions;

(d) expectations of individuals are independent from period to period;

(e) primary buyers and sellers each control only a small part of the total physical resource;

(f) traders are also relatively small;

(g) there is discrimination between primary agents;

we are able to derive simple formulae for the equilibrium number of wet deals, the number of paper deals, the length of the 'daisy chain' and the average proportion of potential traders who are in the market at any one time.

The mechanics of the process are simple: with a given range of expectations held by both primary and secondary agents, there is a range of high expected prices (not lower than the actual forward price plus a risk premium factor) at which agents holding such views are willing to sell. There are also those agents with a range of expected prices (not greater than the actual forward price less a risk premium) who are willing to buy. The actual forward price adjusts to clear this market. *Ex ante*, everyone will believe that they have improved their position. *Ex post*, once chance variations in the actual spot price are averaged out, the primary traders will have paid on average more than they were prepared to pay solely for the hedge, i.e. the average paid by a primary buyer (or seller) is $(R+D)/2$ (with $R<D$). This is a result of the discrimination assumption because it effectively penalizes those whose views are most extreme by charging them their full willingness to pay rather than the marginal value R. This extra resource allows the extension of the paper 'speculative' market and could be very important if D is much greater than R.

In any particular realization of the market the situation is of course much more complex – the spot price outcome will not necessarily equal the forward price and the range of expectations on that 'day' may be larger or smaller than the equilibrium values. Variations of this nature will lead to short-run disequilibrium values – a longer or shorter daisy chain, a temporary increase in 'wet' use of the market, etc.

The adjustment mechanisms should also be considered. As the model makes clear, changes in primary quantities, risk premiums or ranges of expectations all lead to new equilibria with larger or smaller numbers of agents supported by the market and a larger or smaller proportion of the physical commodity being traded on the forward market. This raises the standard problems of exit and entry to a market. Given the growth in the use of the spot market as primary agents learn of its potential, the relevant issue at first is that of entry. However, if the demise of BNOC were to lead to less 'spinning' or less spot market trading for other reasons, there would be the problem of exit to consider. We can imagine that before the full potential became apparent to secondary agents, those who were in the market and held sufficiently diverse views to make deals found themselves making 'abnormal' profits on average (i.e. greater than the minimum required to persuade them to keep trying). This would encourage other firms to enter, and indeed there could even be overshooting if this happened rapidly (because new firms typically do not take account of their effect and the effects of firms like them on the market). If overshooting took place then profits would on average be too low and exit would be required. However, the 'signals' in this type of market will be very difficult to interpret – it is in the nature of a speculative market that 50 per cent of players at any one time will lose. Hence a loser can confuse 'bad luck' or 'poor judgement' leading to losses with the more fundamental problem of over-capacity. This ambiguity of signal might produce asymmetric speeds of adjustment for exit and entry. To enter, the firm looks at the industry as a whole for the signal but to exit it might look solely at its own performance (in the former case it would be difficult for it to do anything else). Over-capacity could therefore produce slow adjustment and, given that the loss makers will be evenly spread as regards their expectations, it would thus tend to extend the length of the daisy chain, (with many firms in the market, there will always be many pairs such that both think that they can do a profitable deal). Hence, if the mechanistic or institutional framework is sensitive to the length of daisy chain, a period of contraction could be a difficult time.

Something can also be said about variability in the lengths of daisy chains. We have argued that in equilibrium the length of the *average* chain will be:

$$C = \frac{R+D}{2r} + 1 \tag{3.27}$$

but some chains will be longer and some shorter than this. Given the distribution of beliefs, all lengths from zero to infinity are possible (although institutional and practical factors will place a sharp upper limit in reality). It is in fact possible to derive the distribution of chain lengths analytically (given the distributions of beliefs) but there seems to be no insight to be gained from what would be a complicated exercise even in the simplest case.

The comparative statics of the model show that the daisy chain length does not respond to the diversity of views of traders but that it does respond to the diversity of views of primary agents. This is because the physical 'base' supplied by primary agents will support a given number of deals: if the range of views of traders changes so that fewer deals can be made, then the larger margins will attract extra traders until the margin is forced down with the number of deals being constant. Hence the size and nature of the paper market is dominated by characteristics of the primary market. The risk premium of traders serves to define the margin that each deal can support in equilibrium.

An increase in D has the effect not only of putting more of Q onto the market but also of putting some on at higher prices (and taking some off at lower prices), thus increasing the maximum surplus that can be extracted from the market. Similarly, changes in R affect both the quantity of hedged primary trade and the margins involved.

ANNEX 4

THE BRENT MARKET AND THE EFFICIENT MARKETS HYPOTHESIS

In a market with forward prices in which agents are able and willing to switch between spot and forward deals, the forward market is seen as a method of reducing risk (while maintaining an adequate return against opportunity cost). This situation envisages buyers of oil (and sellers of oil) debating whether to buy oil (say) one month forward at a certain price or to wait and then buy spot at a price which is at present uncertain. The buyers and sellers will of course have expectations of the future spot price but, since this is uncertain, they will be prepared to trade risk reduction against the price. This trade-off is asymmetric for the two types of agents. Given an expected price P^e, the seller is willing to sell for less now (for the expected future price *minus* the risk premium) in order to have certainty – hence the sellers will be willing to sell now at prices above $P^e - r$, where r is the risk premium for sellers. The buyers view the cost of the transaction in the same way and are therefore willing to pay a price higher than the expected future spot price in order to offset the uncertainty. They are prepared to pay up to $P^e + R$, where R is the buyers' risk premium. This formulation makes it clear that there is a range of prices at which a given buyer and seller could both gain and would hence trade. Even if there is a range of expectations on both sides of the market (as we explain elsewhere), this indeterminacy remains – the actual price established will however bear a relation to P^e. In particular, it is usual to argue that the spot price expected for the next period will be related to the forward price ruling now, i.e:

$$P_t^e(t+1) = a + P_t(t+1) \tag{4.1}$$

where $P_t^e(t+1)$ is the spot price expected at time t to be ruling at time $t+1$. With a forecasting regime that makes random (not systematic) errors this can be replaced by the equation:

$$P_{t+1}(t+1) = a + P_t(t+1) + u_t \tag{4.2}$$

where $E(u_t) = 0$ (u_t being a random variable).

This formulation enables us to test the forecasting ability and market operation for Brent. If the desire to buy forward is motivated solely by the wish to hedge against future uncertain spot prices *and if* agents can forecast correctly *on average*, then, in the regression of the spot price for period $t+1$ on the forward price for that period established in period t, the coefficient should not be significantly different from unity. This is a weak test in some ways because failure can arise from two sources:

(a) biased forecasting (on average under- or over-estimating)
(b) a failure of arbitrage between spot and forward markets

Table A4.1 presents the results of a regression of this type using monthly average Brent data on a 'for the month' basis.

Table A4.1: Regression of Spot Prices on Lagged Forward Prices (monthly data with one-month lag, October 1983–February 1985).

Constant	Lagged Forward Price	R^2	DWS	SEE
6.18	0.78	0.59	1.36	0.73
(4.81)	(0.16)			

The result is striking – the coefficient of the lagged forward price is only 0.78 – but with a standard error of 0.16 this is not quite significantly less than unity. The hypothesis of unbiased forecasting and full intertemporal arbitrage is not rejected by this test.

More powerful tests of this approach can be constructed by analysing the forward price in the spirit of 'rational expectations'. If the forward price is one that will matter in the next period and is the subject of choice relative to the future (unknown) spot price, then the forward price should encapsulate all systematic knowledge of that future price. Any further information on the future price should cause agents to change their pattern of buying forward, and hence the forward price should react accordingly. Hence the forward price should be the best and a sufficient predictor. No other series available at time t should be able to improve on the predictive power of the current forward price for the future spot price. This is not to say that the forward price has to be a very accurate predictor – there may always be a large amount of 'news' after it is settled. In particular, two other prices should not be able to improve on its performance:

— earlier determined forward prices of longer horizons for the same terminal date
— current or earlier spot prices

In Table A4.2 we add the two-month forward price to the regression already presented. This relates the spot price in a period to the one-month forward price from the previous period and the two-month forward price from two periods before. Since the one-month lag has all the information available two months before and also some more recent information, it should be the dominant variable and the two-month forward price should not be significantly different from zero.

The results of this test strongly confirm that the one-month forward price contains all the predictive power of the two-month forward price – there is no systematic relationship of the two-month forward price to the current spot price that is not incorporated by the one-month price. This represents further confirmation of the hypothesis of intertemporal arbitrage and it does not rely on the forecast being unbiased (but rather on the degree of bias being constant).

Table A4.2: Regression of Current Brent Spot Prices on One- and Two-month
Forward Prices (January 1984–February 1985).

Constant	One-month Forward	Two-month Forward	R^2	DWS	SEE
13.72	1.05	−0.53	0.61	1.65	0.77
(7.70)	(0.29)	(0.38)			

However, a final and even more powerful test is obtained when we regress the
current spot price on the one-month lagged spot price and the one-month forward
price of the previous month. The result is shown in Table A4.3.

Table A4.3: Regression of Spot Price on Lagged Spot Price and the Forward Price
of the Previous Month (November 1983–February 1985).

Constant	Lagged Forward Price	Lagged Spot Price	R^2	DWS	SEE
6.21	−0.63	1.41	0.68	1.83	0.65
(4.59)	(0.60)	(0.59)			

The result of this regression is very striking: as with studies on other markets,[1]
which have shown the lagged spot price to have additional forecasting power over
the forward price, we find that for Brent crude the lagged spot price is the
dominant forecasting variable and that the forward price becomes negative and
insignificantly different from zero. The interpretation of this result is that actual
prices follow an autoregressive process so that lagged and current spot values are
correlated. The forward price, which is strongly correlated with the spot price set at
the same time (as we showed above), picks up only some of the variations in the
future price that the autoregression would predict. The forward price, in other
words, is not being used solely for purposes of arbitrage against the future spot
price – if it were it could be set equal to the current spot price and would be more
successful. A similar result is obtained by regressing the current spot on the spot
lagged two months and the forward price as settled two months earlier. The
coefficient of the forward price is negative and insignificant. Alternatively, adding

[1] Tests in the foreign exchange market by other authors (e.g. Hansen and Hodrich in
Journal of Political Economy, 1980, MacDonald in *JPE*, 1983) have similarly rejected the
efficient markets hypothesis. Bird, in Michelsen Institute Paper, CMI–no. 842050–6, rejects
the hypothesis for the Rotterdam spot market for gas oil. Unpublished work by the Institute
on the Nymex futures market for WTI suggests that it too is inefficient. Work by Taylor, in
Streit ed. *Futures Markets* suggests that the markets for cocoa and coffee are possibly efficient.
All of this contrasts with evidence from the stock market where generally it has been
concluded, e.g. Fama in *Foundations of Finance*, that the market is efficient.

both one- and two-month forward prices to the one-month lagged spot price again finds the spot price to be the only significant variable. All of these results point to the use of the forward market for reasons other than physical arbitrage and hedging. The third option for traders is to speculate on this market. However, for speculation to have the possibility of being successful, the expected holding time of the deal (or open position) must be less than that of the degree of forwardness. If the trader did not expect to buy or sell again before the next period's spot price became operative, then the act of speculation, since it is geared to making a margin on the buy/sell transaction, would bring the forward price into line with the expected future spot price. This points to the existence of agents who are willing to buy or sell paper with one or two (or even three) months to run but who expect to have made the opposite deal before delivery takes place, so that they need neither provide nor receive crude oil, and also to have made a profit. Their expected period of holding an open position would be considerably less than the length of the forward contract. Of course it is not possible for all the dealing to be of this nature: as the delivery date on an outstanding contract approaches, the requirement to 'close' the position will force the agent to go much shorter, and the awareness that this can happen (i.e. that an opportunity for profit may not arise in the meantime) means that the speculator must also bear in mind the future spot price that is expected to rule when the contract matures. Thus he is interested in the whole sequence of expected prices from now until maturity, and it is this that keeps the forward price tied to some expected future prices and hence to actual future spot prices.

This tends to characterize the market as having forward prices strongly related to future spot prices, but not picking up all the variation of such prices – or alternatively exhibiting variations not shown in spot prices. At the same time, current spot prices forecast future spot prices very accurately on average, and certainly no residual variation in spot prices that is unexplained by past spot prices can be 'explained' by the forward price from the earlier date.

These results do not conflict with the idea that, on average, spot and forward prices tend to have very similar average values. It just indicates that whatever produces *variations* in spot prices is likely to be reflected in near future spot prices but not so clearly in forward prices. The latter are probably dominated by market pressures that are related to more short-run perceptions of market movements. Since we know that the average daisy chain is in excess of ten links, this confirms that the average 'open' position is only a few days (the average forward contract being perhaps one to one-and-a-half months as shown by our index of 'forwardness' in Chapter 13). All this points to the forward market being dominated by deals that are made with considerations other than expectations of what the price will actually turn out to be in the delivery period.

ANNEX 5

STATISTICAL TESTING AND NORTH SEA CRUDE PRICES

This annex reports the main econometric work lying behind the various pieces of analysis presented in Chapter 14.

The data used in our econometric work are presented in the statistical appendix (Annex 8).

A5.1 The Spot and Term Prices of Brent

We use a monthly series of spot prices for Brent obtained by splicing an industry source available from January 1980 to the *Petroleum Argus* data which start in July 1982.

The Brent spot crude price (PRBREN) is regressed on a constant and a time trend (January 1980–February 1985) with the result shown in equation (5.1):

$$\text{PRBREN} = \underset{(0.54)}{38.42} - \underset{(0.015)}{0.18} \text{ TREND} \tag{5.1}$$

$$\text{SEE} = 2.08, \quad \text{DWS} = 0.78$$

(Standard errors are shown in parentheses)

The trend is significantly negative indicating an average decline of 18 cents per month since January 1980.

Next the Brent term price (BOP) is regressed against a time trend for a similar period (January 1980–March 1985) with the results shown in equation (5.2).

$$\text{BOP} = \underset{(0.51)}{37.49} - \underset{(0.014)}{0.15} \text{ TREND} \tag{5.2}$$

$$\text{SEE} = 2.00, \quad \text{DWS} = 0.36$$

Here the significant negative trend indicates a decline of 15 cents per month on average.

The two prices are related by regressing the differential (spot less official) on a constant over the period July 1982–February 1985 (before July 1982 the differential was more variable so we tried to keep to a fairly homogeneous period). The result (for the period July 1982–February 1985) is shown in equation (5.3).

$$\text{PRBREN} - \text{BOP} = -0.63 \tag{5.3}$$
$$(0.19)$$

$$\text{SEE} = 1.08, \quad \text{DWS} = 0.74$$

Over the shorter data period the Brent official crude price is, on average, 63 cents higher than the Brent spot crude price.

Finally, we include a trend in the equation for the differential and this gives:

$$\text{PRBREN} - \text{BOP} = 0.97 - 0.037 \text{ TREND} \tag{5.4}$$
$$(0.62) \quad (0.017)$$

$$\text{SEE} = 2.43, \quad \text{DWS} = 0.74$$

The significant coefficient on the TREND variable indicates that the Brent spot/term differential has narrowed at an average rate of 3.7 cents per month over the period January 1980–February 1985.

A5.2   The Relationships Between Brent and Other North Sea Crude Prices

The monthly spot price of Brent is regressed on those of Ekofisk and Flotta in models excluding a constant.

The relationship between Brent and Ekofisk spot prices in the period January 1980–January 1985 is given by equation (5.5).

$$\text{PRBREN} = 0.985 \text{ PREKOF} \tag{5.5}$$
$$(0.002)$$

$$\text{SEE} = 0.65, \quad \text{DWS} = 1.46, \quad R^2 = 0.973$$

The relationship between Brent and Flotta spot prices in January 1980–February 1985 is as follows:

$$\text{PRBREN} = 1.018 \text{ PRFLOT} \tag{5.6}$$
$$(0.003)$$

$$\text{SEE} = 0.81, \quad \text{DWS} = 0.84, \quad R^2 = 0.957$$

In both cases the price of Brent is very strongly correlated with the price of the other North Sea crude.

The data are next analysed by regressing the differentials on a constant and a time trend.

The regression of the Brent/Ekofisk spot differential against the trend for the period January 1980–January 1985 is given in equation (5.7).

$$\text{PRBREN} - \text{PREKOF} = -1.065 + 0.018 \text{ TREND} \qquad (5.7)$$
$$\phantom{\text{PRBREN} - \text{PREKOF} = }(0.15) \quad\ (0.004)$$

$$\text{SEE} = 0.57, \quad \text{DWS} = 1.82$$

The regression of the Brent/Flotta spot differential against the trend for the period June 1981–February 1985 is given in equation (5.8).

$$\text{PRBREN} - \text{PRFLOT} = 1.854 - 0.028 \text{ TREND} \qquad (5.8)$$
$$\phantom{\text{PRBREN} - \text{PRFLOT} = }(0.35) \quad\ (0.008)$$

In both cases the opposite signs of intercept and trend indicate that the crude prices are moving towards Brent (one from above and the other from below) at a significant monthly rate.

Estimates of the average Brent/Ekofisk and Brent/Flotta price differentials can be obtained by regressing each differential on a constant. The results are given by equations 1 and 2 in Table A5.1.

Table A5.1: Average Price Differentials Between Brent and Other North Sea Crudes

	Crude	*Data Period*	*Constant (SE)*	*SEE ($)*	*95 per cent confidence range ($)*
1.	Ekofisk	Jan 1980–Jan 1985	−0.49 (0.08)	0.66	−1.81 to +0.83
2.	Flotta	Jan 1980–Feb 1985	0.59 (0.10)	0.81	−1.03 to +2.21

The data show that on average the Brent crude price is 49 cents lower than the Ekofisk crude price. Around this average there is considerable variation. The standard error of estimate of 66 cents implies that, at a 95 per cent confidence interval, the actual differential could be expected to be in the range of −$1.81 to +$0.83. In the case of Brent and Flotta the average differential is estimated at 59 cents. The standard error of 81 cents gives a 95 per cent confidence range for the actual differential of −$1.03 to +$2.21.

A5.3 The Relationships Between Brent and non-North Sea Crude Prices

The monthly spot price of Brent is regressed on those of five major non-North Sea crudes in models excluding a constant. The results are presented in Table A5.2.

Table A5.2: Price Relationships Between Brent and Five non-North Sea Crudes

	Independent Variable	Data Period	Coefficient (SE)	R^2	SEE	DWS
1.	Arabian Light	Jan 1980–Jan 1985	1.031	0.942 (0.004)	0.94	0.74
2.	Arabian Heavy	Jul 1980–Jan 1985	1.104	0.830 (0.007)	1.53	0.33
3.	Forcados	Apr 1980–Jan 1985	0.995	0.969 (0.003)	0.66	1.51
4.	Urals	Jan 1980–Feb 1985	1.016	0.939 (0.004)	0.96	0.84
5.	WTI	Nov 1982–Jan 1985	0.984	0.823 (0.004)	0.59	0.54

Note :   The dependent variable is the Brent spot crude price.

The results show that the Brent spot crude price is closely correlated with all five crudes. However, the squared correlations, 0.830 and 0.823, in equations 2 and 5 respectively, suggest that the Arabian Heavy and WTI spot prices are not as closely related to the Brent spot price as the other crudes. The regression coefficients indicate that the Brent spot price is, on average, higher than those of Arabian Light, Arabian Heavy and Urals (by factors of 3.1, 10.4 and 1.6 per cent respectively) but lower than those of Forcados and WTI (by factors of 0.5 and 1.6 per cent respectively). The analysis is now extended to examine the behaviour of crude price differentials.

The differentials between the spot price of Brent and of five major non-North Sea crudes are regressed on a constant and a trend. However, the TREND variable is significant in the equations for Arabian Heavy and Urals only, and these results are shown in Table A5.3.

Table A5.3: Significant Trends in Price Differentials Between Brent and non-North Sea Crudes

	Crude	Period	Constant (SE)	TREND	SEE	DWS
1.	Urals	Jan 1980–Feb 1985	0.91 (0.25)	−0.011 (0.006)	0.95	0.89
2.	Arabian Heavy	Jul 1980–Jan 1985	4.02 (0.46)	−0.028 (0.012)	1.46	0.42

Both trends are significant and of opposite sign to the constant, indicating a narrowing of the differential *vis-à-vis* Brent.

Average spot price differentials between Brent and the five non-North Sea crudes are obtained by regressing each differential on a constant. The results are presented in Table A5.4 for various periods. The periods are chosen so as to exclude episodes when a differential appears to have been subject to abnormal influences.

For the respective data periods the estimates indicate that, on average, the Brent spot crude price exceeded those of Arabian Light, Arabian Heavy and Urals by $0.98, $3.07 and $0.21 respectively. In contrast, the Brent crude price was lower, on average, than those of Forcados and WTI by 5 and 71 cents respectively. There is considerable variation around these averages (as measured by the standard error of estimate), especially for Arabian Light and Arabian Heavy.

To complete the analysis we provide a matrix of correlation coefficients between the seven crude price differentials for the data period common to all (November 1982–January 1985). This is given by Table A5.5.

Table A5.4: Average Price Differentials Between Brent and non-North Sea Crudes

	Crude	Data Period	Constant (SE)	SEE ($)	95 per cent confidence range ($)
1.	Arabian Light	Jan 1980–Jan 1985	0.98 (0.12)	0.95	−0.92 to +2.88
2.	Arabian Heavy	Jul 1980–Jan 1985	3.07 (0.20)	1.51	+0.05 to +6.09
3.	Forcados	Dec 1982–Jan 1985	−0.05 (0.09)	0.46	−0.97 to +0.87
4.	Urals	Feb 1983–Feb 1985	0.21 (0.13)	0.64	−1.07 to +1.49
5.	WTI	Mar 1983–Nov 1984	−0.71 (0.06)	0.26	−1.23 to −0.19

Table A5.5: Correlations of Spot Crude Price Differentials (November 1982–January 1985)

Brent/	Ekofisk	Flotta	Arabian Light	Arabian Heavy	Forcados	Urals	WTI
Ekofisk	1.000						
Flotta	−0.438	1.000					
Arabian Light	−0.472	0.837	1.000				
Arabian Heavy	−0.627	0.898	0.906	1.000			
Forcados	0.470	0.212	0.095	0.012	1.000		
Urals	−0.105	0.735	0.742	0.598	0.497	1.000	
WTI	0.407	−0.547	−0.754	−0.684	−0.047	−0.507	1.000

ANNEX 6

ANALYSIS OF THE RELATIONSHIP BETWEEN THE BRENT NETBACK AND SPOT CRUDE PRICE

A6.1 Introduction

This paper sets out to examine the nature of the relationship between the Brent crude price and the netback obtained from using Brent to produce products. In particular, we are interested to see whether the margin between the two stays constant on average or undergoes substantial shifts, and, if it does, to see whether we can explain why the shifts take place.

To carry out this exercise we require data on prices and netbacks over a lengthy period. Published sources do not go back far enough and so we are forced to 'splice' together data from different sources. Data on netbacks, based on complex refining at Rotterdam, have been given to us by an industry source for the period January 1981 to December 1983, together with the company's assessment of the spot crude price of Brent for the same period. For the period September 1983 to February 1985 we use published data on the Brent netback (using complex refining at Rotterdam) from *PIW* and on Brent spot crude prices from *Petroleum Argus*. We find the average differential between sources for the four months common to both series and then adjust the 1984–5 data by these differentials. The constructed series run from January 1981 to February 1985. There are, therefore, fifty monthly observations.

A6.2 The Normal Margin Model

The various tests we carry out on the relationship between the two series are based on those used in an earlier working paper by the Institute which analysed the price structures of Arabian Light and Nigerian Light.[1] The first test is to look to see whether there is a constant dollar for dollar relationship *at the margin* between the Brent netback BN_t and the Brent spot crude price BS_t. This is the 'normal margin' model and it can be specified as:

$$BN_t = a + BS_t + u_t \qquad (6.1)$$

where u_t is a random term at time t and the expected value of u_t is zero. The normal margin model is first estimated for the whole of the data period and the results are presented in equation 1 of Table A6.1.

[1] Bacon, R.W. *A Study of the Relationship Between Spot Product Prices and Spot Crude Prices*, Working Paper WPM5, Oxford Institute for Energy Studies, 1984.

Table A6.1: Estimated Brent Normal Margins for January 1981–February 1985
and for Sub-periods

Equation Number	Dependent Variable	Data Period	Constant (SE)	SEE	DWS
1.	$BN_t - BS_t$	Jan 1981– Feb 1985	1.58 (0.190)	1.34	0.57
2.	$BN_t - BS_t$	May 1981– Jan 1983	2.87 (0.186)	0.85	1.76
3.	$BN_t - BS_t$	Feb 1983– Feb 1985	0.69 (0.139)	0.70	0.96

The estimates show that, on average, the netback value is $1.58 *higher* than the spot crude price over the data period. There is a large variation around this normal margin. The standard error of estimate ($1.34) suggests that for a 95 per cent confidence level, the actual margin is expected to be in the range −$1.10 to +$4.26. Clearly, the range of actual netback margins that is consistent with the long-run normal margin hypothesis is too wide to be of much practical relevance. In addition, the low Durbin-Watson statistic (0.57) implies misspecification of the relationship between Brent netbacks and spot crude prices. Some important and systematic factors are apparently ignored by the normal margin model. This assertion is supported by examination of a plot of the actual values, centred on the estimated normal margin (see Figure A6.1). There appears to be a period of relatively high netback margins from around May 1981 to January 1983. This is followed by an episode of much lower netback margins (February 1983–February 1985). In order to assess the structural breaks the data are separated into these two sub-periods. Equations 2 and 3 in Table A6.1 estimate the normal margin for each sub-period.

The results provide evidence of a structural break in the middle of the data period. The first sub-period shows a positive normal margin of $2.87 with a standard error of estimate of 85 cents. This implies a 95 per cent confidence range for the actual margin of +$1.17 to +$4.57. Hence, although the standard error of estimate is much reduced, the range of actual netback margins remains too wide for unqualified acceptance of the normal margin hypothesis. However, the high Durbin-Watson statistic (1.76) suggests a much improved performance by the normal margin model for this sub-period relative to the whole period. The estimates for the second sub-period give a positive netback margin of 69 cents (a fall in the margin of more than $2 between sub-periods). The standard error of estimate is now 70 cents. The 95 per cent confidence range for the actual margin becomes −$0.71 to +$2.09. The range is again large and, coupled with model misspecification (as judged by the Durbin-Watson statistic), we must reject the normal margin hypothesis.

In summary, the normal margin model appears to provide an inadequate representation of the relationship between Brent netbacks and spot crude prices. This is the case for the whole of our data period and for identified sub-periods.

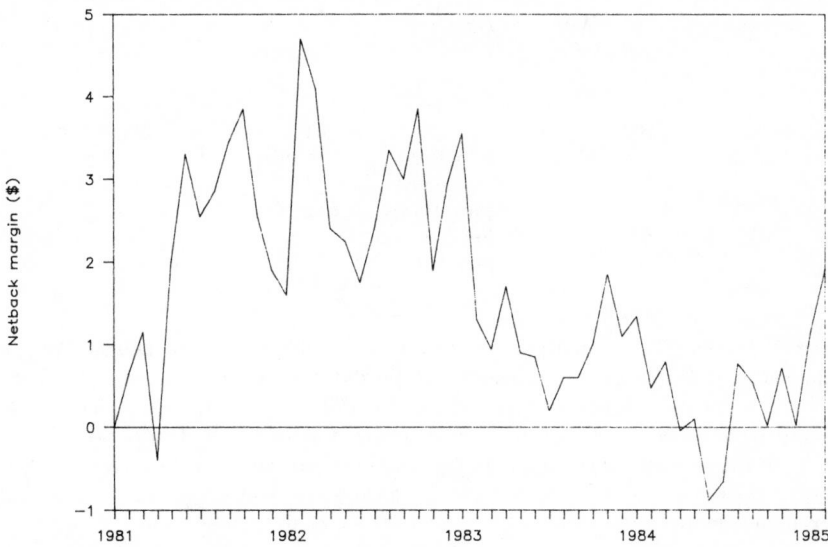

Figure A6.1   Brent Netback Margin Movement. January 1981–February 1985

However the data strongly suggest that an important shift in the relationship between the two series took place in early 1983.

A6.3   Brent Netbacks and Spot Crude Prices: An Unconstrained Relation

We drop the assumption of a dollar for dollar relation between the Brent netback and the spot crude price. Instead, we estimate an unconstrained relationship which allows movements other than dollar for dollar responses. This can be specified thus:

$$BN_t = a + bBS_t + u_t \tag{6.2}$$

This model is estimated for the whole data period as well as for the two sub-periods. The results are presented in Table A6.2. The constant is generally insignificant so we report the regressions estimated without a constant.

For the period as a whole, the slope coefficient implies that a $1 increase in the Brent spot crude price is associated with a $1.05 rise in the Brent netback. However, the adoption of a 'spot-related' margin does little to reduce the instability of our estimate with only a marginal improvement in the standard error of estimate. In addition, there is very little change in the Durbin-Watson statistic, which again indicates serious misspecification of the model. The results for the sub-period May 1981–January 1983 show a slight deterioration relative to the normal

Table A6.2: Relations Between Brent Netbacks and Spot Crude Prices

Equation Number	Dependent Variable	Data Period	Spot Price Coefficient (SE)	SEE	DWS
1.	BN_t	Jan 1981– Feb 1985	1.05 (0.006)	1.32	0.66
2.	BN_t	May 1981– Jan 1983	1.08 (0.006)	0.99	1.67
3.	BN_t	Feb 1983– Feb 1985	1.02 (0.005)	0.70	0.95

margin model. The standard error of estimate is higher while the Durbin-Watson statistic falls slightly. For the second sub-period, the overall performance of the model is almost unchanged relative to the normal margin model. The slope coefficients for both periods are similar to that for the whole data period.

Both the normal margin and 'spot-related' margin models are characterized by unstable estimates and problems of omitted variables. This suggests that we should examine whether other relevant variables influence the relationships.

A6.4   A Revised Model with the Brent Official Price

In the Institute's earlier work on Arabian Light and Nigerian Light we found that the margin between the netback and the spot price shifts significantly once the spot price falls below the official price (in mid-1981).[2] The argument is that, while spot is above official it is clearly in short supply and is marginal, while when spot is lower than official it is in abundant supply and is still the marginal source of supply, but in this case refiners will wish to keep future contracts by buying a certain amount of official even though it is more expensive. Hence, in the latter case, the netback price reflects the costs of all marginal inputs; and so official prices as well as spot prices are found to be related to the netback in the post-1981 period.

The success of our tests for Arabian Light and Nigerian Light suggests that we should examine the Brent data to see whether the two-tier pricing system is reflected at all by the netbacks.

We therefore estimate the equation:

$$BN_t = f(BS_t, BO_t) \tag{6.3}$$

where BO_t is the Brent official crude price

The Brent netback should be related to a weighted average of spot and official crude prices whenever the spot price falls *below* the official price. Over the whole data period the spot price is below the official price in thirty-one of the fifty

[2] *ibid.*

months. The two crude prices are equal in four months and spot is above official in the remaining fifteen months. Therefore, we might expect to find a significant coefficient on the official price over the whole period. As for the sub-periods, the second sub-period (February 1983–February 1985) shows a greater tendency than the first sub-period (May 1981–January 1983) for the spot price to be below the official price. Hence, the official price should perform more strongly in the netback equation for the second sub-period. The results are given in Table A6.3.

Table A6.3: The Impact of Spot and Official Prices on Brent Netbacks

Equation Number	Dependent Variable	Data Period	Spot Coefficient (SE)	Official Coefficient (SE)	SEE	DWS
1.	BN_t	Jan 1981–Feb 1985	0.85* (0.114)	0.19** (0.112)	1.29	0.52
2.	BN_t^c	Jan 1981–Feb 1985	0.67* (0.080)	0.35* (0.080)	0.85	2.08
3.	BN_t	May 1981–Jan 1983	0.88* (0.085)	0.20* (0.084)	0.88	1.58
4.	BN_t^c	May 1981–Jan 1983	0.79* (0.085)	0.29* (0.083)	0.79	2.16
5.	BN_t	Feb 1983–Feb 1985	0.79* (0.154)	0.23 (0.151)	0.69	0.96
6.	BN_t^c	Feb 1983–Feb 1985	0.77* (0.188)	0.25 (0.185)	0.61	1.78

Notes: * denotes significance at 5 per cent confidence level
       ** denotes significance at 10 per cent confidence level
        c denotes estimation by Cochrane-Orcutt technique

For the period as a whole, the coefficient on the official price is statistically significant as expected. In addition, the goodness of fit is improved somewhat when the equation is estimated by the Cochrane-Orcutt technique in order to control for serial correlation. The presence of serial correlation suggests some systematic specification error. The results for the first sub-period also provide evidence for a positive association between the official price and the netback. For this sub-period the Cochrane-Orcutt estimation technique improves the goodness of fit only marginally. The estimates for the second sub-period indicate no relationship between the Brent netback and official crude prices.

The problem with this analysis is that, although the sub-periods roughly correspond to episodes where the official price is below and then above the spot price, the relationship is not exact. Hence we divide the data on a different basis, characterizing them by the criterion of whether the official is above or below the spot. The results are shown in Table A6.4. This new division of the data fits much better to our hypothesis. When spot is above official it is the marginal input and official prices are not reflected in the netback, but when spot is below official both are reflected in the netback to a significant extent. There is still however a suggestion of specification error from the Durbin-Watson statistic.

Table A6.4: The Impact of Spot and Official Prices on Brent Netbacks

Equation Number	Dependent Variable	Data Length	Spot Coefficient (SE)	Official Coefficient (SE)	SEE	DWS
(a) Periods when spot greater than official						
1.	BN_t	15 months	1.05* (0.011)		1.34	0.64
2.	BN_t	15 months	1.27* (0.39)	−0.23 (0.40)	1.37	0.68
(b) Periods when spot less than official						
3.	BN_t	31 months	1.05* (0.008)		1.35	0.85
4.	BN_t	31 months	0.54* (0.16)	0.49* (0.15)	1.19	0.77

A6.5　The Dynamic Equations

Even though we are using monthly average data there is a possibility that there are lagged relationships between spot crude prices and netbacks.　We represent the general lag structure by dynamic supply and demand equations:

$$BN_t = f(BS_t, BS_{t-1}, ...) \qquad (6.4)$$

and:

$$BS_t = g(BN_t, BN_{t-1}, ...) \qquad (6.5)$$

In the two-tier price system, this representation is unaltered if the spot price exceeds the official price. However, when the spot price is below the official price we might expect both lagged crude prices to affect the netback. The dynamic equations are estimated for the period as a whole as well as for each sub-period. Table A6.5 presents the estimates of the supply equations.

The overall picture is that the lagged spot crude price adds nothing to an understanding of the behaviour of the Brent netback. The goodness of fit of the equations is virtually unchanged. The lagged spot price is only significant (even at the 10 per cent confidence level) for the sub-period May 1981–January 1983 when the official price is omitted. The official price again shows up well when we correct for serial correlation by using the Cochrane-Orcutt estimation technique. This even applies to the second sub-period for which the coefficient on the official price

Table A6.5: Lags Between Brent Netbacks and Spot Crude Prices

Equation Number	Dependent Variable	Data Period	Spot	Lagged Spot	Official	SEE	DWS
1.	BN_t	Jan 1981–Feb 1985	0.86* (0.121)	0.05 (0.182)	0.14 (0.179)	1.29	0.55
2.	BN_t	Jan 1981–Feb 1985	0.89* (0.114)	0.16 (0.113)		1.29	0.58
3.	BN_t^c	Jan 1981–Feb 1985	0.69* (0.083)	−0.005 (0.028)	0.36* (0.089)	0.84	2.06
4.	BN_t	May 1981–Jan 1983	0.88* (0.090)	−0.01 (0.156)	0.21 (0.158)	0.91	1.60
5.	BN_t	May 1981–Jan 1983	0.92* (0.087)	0.16** (0.086)		0.93	1.49
6.	BN_t^c	May 1981–Jan 1983	0.78* (0.087)	0.03 (0.138)	0.27** (0.155)	0.81	2.11
7.	BN_t	Feb 1983–Feb 1985	0.95* (0.207)	−0.25 (0.214)	0.32** (0.168)	0.68	1.03
8.	BN_t	Feb 1983–Feb 1985	1.09* (0.203)	−0.07 (0.203)		0.72	0.97
9.	BN_t^c	Feb 1983–Feb 1985	0.89* (0.211)	−0.22 (0.192)	0.35** (0.204)	0.61	1.90

Notes: * denotes significance at 5 per cent confidence level
   ** denotes significance at 10 per cent confidence level
   Standard errors in parentheses
   c denotes estimation by Cochrane-Orcutt technique

Table A6.6: Lags Between Brent Spot Crude Prices and Netbacks

Equation Number	Dependent Variable	Data Period	Netback	Lagged Netback	Official	SEE	DWS
1.	BS_t	Jan 1981–Feb 1985	0.82* (0.118)	−0.30* (0.139)	0.44* (0.100)	1.06	1.23
2.	BS_t	Jan 1981–Feb 1985	0.90* (0.137)	0.05 (0.136)		1.25	0.67
3.	BS_t^c	Jan 1981–Feb 1985	0.82* (0.100)	0.003 (0.040)	0.13 (0.111)	0.93	1.91
4.	BS_t	May 1981–Jan 1983	1.02* (0.126)	−0.13 (0.180)	0.03 (0.146)	0.94	1.69
5.	BS_t	May 1981–Jan 1983	1.03* (0.123)	−0.10 (0.122)		0.91	1.70
6.	BS_t^c	May 1981–Jan 1983	1.04* (0.129)	−0.09 (0.182)	−0.04 (0.160)	0.91	1.93
7.	BS_t	Feb 1983–Feb 1985	0.61* (0.153)	0.11 (0.140)	0.25** (0.149)	0.64	0.72
8.	BS_t	Feb 1983–Feb 1985	0.76* (0.131)	0.22** (0.130)		0.66	0.70
9.	BS_t^c	Feb 1983–Feb 1985	0.47* (0.119)	0.30* (0.098)	0.20 (0.150)	0.44	1.61

Notes: *   denotes significance at 5 per cent confidence level
        ** denotes significance at 10 per cent confidence level
             Standard errors in parentheses
         c   denotes estimation by Cochrane-Orcutt technique

is significant at the 10 per cent confidence level when previously insignificant. The Cochrane-Orcutt estimation technique also brings slight improvements in the goodness of fit of the equations.

The estimates of the dynamic demand equations are given in Table A6.6. The evidence for a lag structure on the demand side is also not very convincing. For the whole period, there is a significant coefficient on the lagged netback (equation 1) but it possesses the wrong sign. In any case its significance disappears once we correct for serial correlation. The only evidence for a lagged effect on the demand side is contained in the second sub-period (February 1983–February 1985). There is a stronger tendency for spot prices to be below official prices in the second sub-period. Hence, there may be short lags on the demand side when there is excess supply (spot below official) but not when there is excess demand (spot above official). Of course, we also have to accept that shifts in supply are immediately translated into changes in product prices whether or not there is excess supply.

A6.6   Conclusions

The simple data on the netback margin for complex refining based at Rotterdam for Brent show that over the whole of the period 1981–5 the margin is positive (unlike the case for refining of Arabian Light that we had investigated before). However the margin shows not only substantial variability between successive months but also a very clear decline in average value from February 1983 onwards (the first four months of 1981 are also very low). Tests on the data suggest that the average margin shifts by a significant amount between the two periods. However, statistical analysis also suggests that some other factor is systematically affecting netbacks (which is not reflected in spot prices). We test to see whether the 'official' Brent price also has an influence on netbacks by its possible role as a joint input together with spot crude. The evidence here lends support to the view that official prices are moderating the relationship between spot crude and netbacks in periods when the spot price is below the official price.

There is virtually no evidence that the variability of the margin is caused by the existence of lagged relationships between the two magnitudes.

Clearly we have an incomplete account of the shifts in the netback margin. A particular problem for this analysis is that the need to splice data sources together means that there may be inaccuracies in the data. A special problem of which we are aware is the fact that the spot price data used do not distinguish between current and forward prices for the pre-1984 period. The amalgamation of the two may bring some changes that would help to give a clearer link to netbacks.

ANNEX 7

THE SHORT-RUN SPOT PRICE VARIABILITY OF CERTAIN CRUDES

A7.1 Introduction

From certain *a priori* reasoning about the structure of the oil market we may expect that some crudes will have prices that are more variable than those of others. This annex provides tests on six crudes:

— Brent
— WTI
— Arabian Light
— Forcados
— Ninian
— Dubai

The period of testing is from 1st January 1984 to end February 1985 using daily data from *Platt's Crude Oil Marketwire*. The Forcados and Ninian series begin in July 1984 and so can be tested only for a sub-period. At the maximum we have 290 observations.

A7.2 The Tests

In testing short-run variability we have to face the problem that there are of course some changes in the price that persist in the medium run. If differentials change then one series may vary more than the other for these structural reasons. Hence it is important to carry out tests for periods or sub-periods in which there are no major trends (or cycles) in the data.

We check this for the period as a whole in two ways. We carry out a regression of price on a constant and a linear trend (not for Forcados and Ninian because of the shorter period). The results are shown in Table A7.1.

These results show that the price dropped slowly but significantly on average through the period (at 1.1 cents per trading day on average for Brent, etc.) for all four crudes. Hence there is some medium-term variability in the series to be removed before tests on short-run variability can be carried out. The picture is made even more conclusive by the very low Durbin-Watson statistic which indicates series misspecification. This pattern in the trend corrected residuals means that some other factor may be present which creates medium-term

Table A7.1: Trends in Crude Prices. January 1984–February 1985.

Crude	Constant	Trend	SEE	DWS
Brent	30.23	−0.011	0.65	0.09
	(0.07)	(0.0004)		
WTI	31.27	−0.016	0.86	0.07
	(0.10)	(0.0006)		
Arabian Light	28.57	−0.004	0.31	0.12
	(0.04)	(0.0002)		
Dubai	27.97	−0.003	0.32	0.14
	(0.04)	(0.0002)		

(Standard errors in parentheses)

variations in the errors. This must also be removed before an estimate of short-run variability is made.

The second way in which the data are analysed is simply to plot the four series against time. The graphs of 290 observations are too large to be usefully reproduced here but all four showed clear structural shifts. There is an episode of 'higher' prices until around observation 120 (20th June). This is followed by an episode of 'medium' prices until around observation 220 (13th November). Finally there is an episode of 'lower' prices until observation 270 (30th January 1985) followed by a return to medium prices. For Arabian Light the second and third periods are not distinct.

As a result of these observations the data are divided into three sub-periods (the last twenty days are put into the third period) with the periods being the same for each crude. This means that the optimal structural breaks are not identified since they appear to be slightly different for each crude. However, the advantages of a uniform treatment greatly outweigh the loss of precision this entails.

For each sub-period the error variance of each series is calculated (relative to the period mean) by regression on a constant. This yields the Durbin-Watson statistic which is a useful check in this situation. The results are given in Table A7.2. These results are quite strong:

(a) In all periods Arabian Light and Dubai show the least variation (ranging from an average of 14 cents per day in early 1984 to 33 cents per day in mid-1984).

(b) Forcados occupies a middle position and is much more variable than Arabian Light and Dubai – around 50 cents per day from July 1984 onwards.

(c) Brent and Ninian are more variable still – up to 67–69 cents per day from November 1984 onwards.

(d) WTI is the most variable in general but is the same as Brent and Ninian in mid-1984. By the end 1984 to early 1985 period the average variation is almost $1 per day around the mean for the period.

Table A7.2: Standard Errors of Estimate for Various Crudes for Certain Sub-
periods

Crude	Period	SEE ($)	DWS
Brent	Jan 1–Jun 20 1984	0.33	0.17
WTI	Jan 1–Jun 20 1984	0.41	0.13
Arabian Light	Jan 1–Jun 20 1984	0.14	0.16
Dubai	Jan 1–Jun 20 1984	0.19	0.09
Brent	Jul 5–Nov 13 1984	0.59	0.19
WTI	Jul 5–Nov 13 1984	0.57	0.21
Arabian Light	Jul 5–Nov 13 1984	0.33	0.18
Forcados	Jul 5–Nov 13 1984	0.45	0.25
Ninian	Jul 5–Nov 13 1984	0.58	0.21
Dubai	Jul 5–Nov 13 1984	0.27	0.41
Brent	Nov 13 1984–Feb 28 1985	0.67	0.11
WTI	Nov 13 1984–Feb 28 1985	0.94	0.10
Arabian Light	Nov 13 1984–Feb 28 1985	0.22	0.25
Forcados	Nov 13 1984–Feb 28 1985	0.49	0.11
Ninian	Nov 13 1984–Feb 28 1985	0.69	0.11
Dubai	Nov 13 1984–Feb 28 1985	0.32	0.13

(e) The periods are ranked consistently for all crudes apart from Arabian Light –
early 1984 shows the lowest variation and late 1984/early 1985 the highest
variation.

(f) The Durbin-Watson statistics are universally bad but this is no surprise even
with short periods because we have shown elsewhere that prices are very
strongly autoregressive (almost random walks). This means that we could
expect similar orderings in (say) first differences of prices. This would
indicate the extent to which the series is affected by 'news' as opposed to
recent history (as embodied in the day before's price and its memory of the
day before, etc.).

(g) We could apply tests of significance to the variances but these would be of
doubtful value because of the lack of serial independence. The critical value
for an F statistic with 120 and 120 degrees of freedom and a 5 per cent test is
1.35. Hence the ratio of the larger to the smaller SEE squared (given that both
have the same degrees of freedom) would test for the equality of variances
between crudes. Equivalently if the ratio of larger to smaller SEE is greater
than 1.16 then the variances would be significantly different if there were no
serial correlation.

As a final check on variability we difference the data (effectively assuming that
they are generated by a random walk) and repeat the regressions on a constant for
the sub-periods. The results are shown in Table A7.3.

Table A7.3: Standard Errors of Estimate from Random Walk (First Differenced) Models

Crude	Period	SEE ($)	DWS
Brent	Jan 1–Jun 20 1984	0.141	2.12
WTI	Jan 1–Jun 20 1984	0.147	2.15
Arabian Light	Jan 1–Jun 20 1984	0.055	2.31
Dubai	Jan 1–Jun 20 1984	0.058	2.10
Brent	Jun 20–Nov 13 1984	0.224	1.84
WTI	Jun 20–Nov 13 1984	0.256	1.97
Arabian Light	Jun 20–Nov 13 1984	0.139	1.61
Forcados	Jun 29–Nov 13 1984	0.211	2.08
Ninian	July 2–Nov 13 1984	0.264	1.85
Dubai	Jun 20–Nov 13 1984	0.167	1.88
Brent	Nov 13 1984–Feb 28 1985	0.228	1.58
WTI	Nov 13 1984–Feb 28 1985	0.301	1.94
Arabian Light	Nov 13 1984–Feb 28 1985	0.114	1.58
Forcados	Nov 13 1984–Feb 28 1985	0.162	1.83
Ninian	Nov 13 1984–Feb 28 1985	0.231	1.61
Dubai	Nov 13 1984–Feb 28 1985	0.117	1.81

These results confirm the previous findings. The rankings are absolutely consistent.

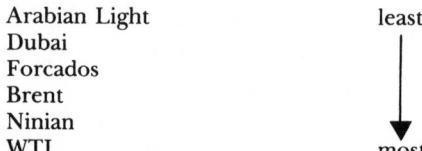

In terms of periods the picture is of the first period as the most stable with the second and third not very different.

Significance tests between crudes show that:

(a) For the first period: Brent and WTI are equally variable but both significantly more so than Arabian Light and Dubai. Arabian Light and Dubai are equally variable.

(b) For the second period: Ninian and WTI are equally variable but Ninian is more variable than Brent. Moreover, Ninian and WTI are more variable than Forcados. These four crudes are all more variable than Arabian Light and Dubai. Dubai is more variable than Arabian Light.

(c) For the third period: (the critical F value at 80, 80 – the nearest value tabulated – is 1.44): WTI is significantly more variable than Brent and Ninian which in turn are both more variable than Forcados. All these are more variable than Dubai and Arabian Light which are equally variable.

The consistency of the picture does add extra evidence to support our general ranking. The DWS is now no longer indicating misspecification so that the errors are not interfering with the tests' validity.

In summary, the series for prices which, apart from long-run trends, have been shown to respond to two components:

— history (in terms of yesterday's price)
— news (in terms of a shock on the day in question);

reveal that the greatest short-run variability is for WTI, closely followed by those for Ninian and Brent. Forcados is much more stable and Dubai and Arabian Light more stable still (with perhaps only half the variability of WTI).

A7.3   Short-run Variability

For a series in first differenced form the error variance is a measure of the effect of 'news' on the series. A shock affects the change in the level of the series, which feeds into the next difference, etc. Technically, if the series were a pure random walk then the variance of the series would be infinite. The fact that the variance of the series is not more than ten times that of the differenced series points to the fact that we have established elsewhere: the series are not pure random walks but tend rather to be strongly autoregressive around equilibrium values. It would be possible to estimate the autoregressions, but, given the evidence on Brent and WTI that the coefficients are around 0.99, we feel that results based on first differencing are fully representative.

ANNEX 8

STATISTICAL APPENDIX

Table S.1: UK. Exploration and Development Drilling. 1964–84.

Year	Exploration Wells Started	Appraisal Wells Started	Development Wells Started	Mobile Rig Activity (Rig years)	Fixed Platform Activity (Rig years)	Total Rig Time (Rig years)
1964	1	0	0	0.02	0	0.02
1965	10	0	0	2.6	0	2.6
1966	20	8	3	6.4	0.5	6.9
1967	42	16	13	8.8	2.4	11.2
1968	31	7	36	6.0	5.3	11.3
1969	44	8	27	7.7	4.5	12.2
1970	22	2	28	5.3	3.3	8.6
1971	24	4	34	5.2	3.7	8.9
1972	33	8	36	8.8	3.8	12.6
1973	42	19	21	13.3	3.2	16.5
1974	67	33	20	24.5	2.8	27.3
1975	79	37	21	27.7	2.6	30.3
1976	58	28	54	21.2	9.3	30.5
1977	67	38	96	23.6	14.9	38.5
1978	37	25	96	18.1	18.6	36.7
1979	33	15	102	16.1	21.5	37.6
1980	32	22	122	20.6	25.2	45.8
1981	48	26	137	24.6	27.0	51.6
1982	68	43	118	30.1	25.0	55.1
1983	77	51	95	34.2	24.1	58.3
1984	106	76	108	49.0	27.6	76.6

Source: *Brown Book*, various issues

Table S.2:   Norway. Exploration and Development Drilling. 1966–83.

Year	Exploration Wells Started	Appraisal Wells Started	Development Wells Completed
1966	2	–	–
1967	6	–	–
1968	10	2	–
1969	12	1	–
1970	14	3	2
1971	11	5	1
1972	11	3	2
1973	16	6	4
1974	12	6	19
1975	20	6	33
1976	20	3	11
1977	11	9	18
1978	14	5	19
1979	16	12	104
1980	26	10	49
1981	25	14	15
1982	35	14	n.a.
1983	33	7	n.a.

Sources: Royal Ministry of Petroleum and Energy, Norway, Storting Report No. 53 (1979–80), *Concerning the Activity on the Norwegian Continental Shelf* and *Fact Sheet*, various issues.
Energy Information Administration, US Department of Energy, *The Petroleum Resources of the North Sea*, 1983.

Table S.3: UK. Producing Fields. Discovery and Production Start-up Dates.

Field	Date of Discovery	Field	Date of Production Start-up	Field	Development Lag Years	Development Lag Months
Montrose	Dec 69	Argyll	Jun 75	Argyll	3	10
Forties	Nov 70	Forties	Sep 75	Forties	4	10
Auk	Feb 71	Auk	Dec 75	Auk	4	10
Brent	Jul 71	Beryl A	Jun 76	Beryl A	3	9
Argyll	Aug 71	Montrose	Jun 76	Montrose	6	6
Beryl A	Sep 72	Brent	Nov 76	Brent	5	4
S Cormorant	Sep 72	Piper	Dec 76	Piper	3	11
Deveron	Sep 72	Claymore	Nov 77	Claymore	3	5
Piper	Jan 73	Thistle	Feb 78	Thistle	4	7
Maureen	Feb 73	Dunlin	Aug 78	Dunlin	5	1
Dunlin	Jul 73	Heather	Oct 78	Heather	4	10
Thistle	Jul 73	Ninian	Dec 78	Ninian	4	8
Heather	Dec 73	Statfjord (UK)	Nov 79	Statfjord (UK)	5	7
Hutton	Dec 73	S Cormorant	Dec 79	S Cormorant	7	3
Ninian	Apr 74	Murchison (UK)	Sep 80	Murchison (UK)	5	0
Statfjord (UK)	Apr 74	Tartan	Jan 81	Tartan	6	0
Claymore	Jun 74	Buchan	May 81	Buchan	6	9
Magnus	Jul 74	Beatrice	Sep 81	Beatrice	5	0
Buchan	Aug 74	N Cormorant	Feb 82	N Cormorant	7	6
N Cormorant	Aug 74	Fulmar	Feb 82	Fulmar	6	2
Tartan	Jan 75	NW Hutton	Apr 83	NW Hutton	8	0
NW Hutton	Apr 75	S Brae	Jul 83	S Brae	6	0
Scapa	Aug 75	Magnus	Aug 83	Magnus	9	1
Murchison (UK)	Sep 75	Maureen	Sep 83	Maureen	10	7
Fulmar	Dec 75	Duncan	Nov 83	Duncan	2	10
Highlander	Apr 76	Hutton	Aug 84	Hutton	10	8
Beatrice	Sep 76	Deveron	Sep 84	Deveron	12	0
S Brae	Jul 77	Innes	Jan 85	Innes	1	10
Duncan	Jan 81	Highlander	Feb 85	Highlander	8	10
Innes	Mar 83	Scapa	Sep 85	Scapa	10	1

Source: *Brown Book*, various issues, and Wood Mackenzie.

Table S.4: Norway. Producing Fields. Discovery and Production Start-up Dates.

Field	Date of Discovery	Field	Date of Production Start-up	Development Lag	
				Years	Months
Ekofisk	Dec 69	Ekofisk	Jul 71	1	7
		W Ekofisk	May 77	7	5
		Cod	Dec 77	8	0
		Tor	Jun 78	8	6
		Albuskjell	May 79	9	5
		Eldfisk	Aug 79	9	8
		Edda	Dec 79	10	0
Statfjord	Feb 74	Statfjord	Dec 79	5	10
Valhall	Jun 75	Murchison	Sep 80	5	0
Murchison	Sep 75	Valhall	Oct 82	7	4

Source: Wood Mackenzie

Table S.5: UK. Monthly Production of Crude Oil. 1980–85. Thousand Barrels per Day.

	1980	1981	1982	1983	1984	1985
Jan	1601	1726	1902	2139	2573	2768
Feb	1654	1771	1914	2311	2649	2692
Mar	1680	1851	1961	2249	2514	2623
Apr	1511	1818	2094	2166	2524	2644
May	1617	1705	2079	2230	2486	2621
Jun	1589	1738	2074	2059	2383	2219
Jul	1579	1732	2073	2298	2507	2412
Aug	1532	1796	2052	2313	2339	2241
Sep	1529	1850	2130	2374	2472	2610
Oct	1588	1850	2167	2371	2652	2708
Nov	1706	1907	2138	2491	2642	2673
Dec	1759	1850	2213	2549	2681	2483
Average	1612	1802	2066	2296	2535	2558

Source: Wood Mackenzie

Table S.6: Norway. Monthly Production of Crude Oil. 1980–85. Thousand Barrels per Day.

	1980	1981	1982	1983	1984	1985
Jan	571	537	555	563	730	730
Feb	627	519	545	663	680	735
Mar	563	527	535	655	678	761
Apr	546	500	590	661	690	771
May	548	565	488	645	716	771
Jun	556	570	482	676	725	679
Jul	269	549	539	702	641	847
Aug	538	346	523	601	712	824
Sep	474	518	470	620	750	854
Oct	552	432	431	680	784	893
Nov	533	530	558	707	763	911
Dec	558	531	605	719	800	902
Average	528	509	527	658	722	807

Source: Wood Mackenzie

Table S.7:   UK and Norway. Producing Oilfields and their Licensees. Percentage
Shares. 1984.

Fields, UK	*Licensees*
1. Argyll	Associated News (12.5), Hamilton Oil (36.0), Kleinwort Benson (2.5), RTZ (25.0), Texaco (24.0)
2. Auk	Esso (50.0), Shell (50.0)
3. Beatrice	Britoil (28.0), Deminex (22.0), Hunt (10.0), Kerr McGee (25.0), Lasmo (15.0)
4. Beryl	Amerada Hess (20.0), Enterprise Oil (10.0), Mobil (50.0), Texas Eastern (20.0)
5. Brae	BC Resources (13.1), Bow Valley (4.2), Britoil (20.0), Kerr McGee (8.0), Louisiana Land (10.7), Marathon (38.0), Norsk Hydro (2.0), Sovereign (4.0)
6. Brent	Esso (50.0), Shell (50.0)
7. Buchan	BP (24.58), Charterhall (4.14), Charterhouse (12.71), Clyde (12.71), Goal Pet (4.54), Lochiel (0.90), Sulpetro (12.71), Texaco (9.23), Transworld (12.71), Tricentrol (5.76)
8. Claymore	AB Exploration (0.50), Agip (2.50), Allied (20.00), Anvil (0.50), Coalite (1.00), Dow Chemical (5.00), Floyd (0.50), Getty (21.20), ITO (20.00), North Sea and General (1.00), Occidental (23.40), Pict (0.50), Sovereign (2.00), Texas Gas (0.60), Third Triton (0.50), Transworld (0.80)
9. Cormorant	Esso (50.0), Shell (50.0)
10. Cormorant S	Esso (50.0), Shell (50.0)
11. Duncan	Associated News (12.50), Hamilton Oil (36.0), Kleinwort Benson (2.50), RTZ (25.0), Texaco (24.0)
12. Dunlin	Britoil (9.77), Conoco (9.77), Esso (35.35), Gulf (9.77), Shell (35.35)
13. Forties	BP (83.13), Berkeley Exploration (0.25), Charterhall (0.25), Charterhouse (1.25), CPL (0.25), Elf (1.50), Esso (2.61), Hispanoil (0.25), Ind Scot Energy (0.25), Norsk Hydro (0.25), North Sea and General (0.25), Norwich Union (0.25), OK Exploration (0.95), Plascom (0.25), RTZ (1.00), Saxon (0.25), Shell (2.61), Sovereign (0.50), Texaco (1.00), Trafalgar House (1.45), Ultramar (1.00), Union Jack (0.25), Viva (0.25)
14. Fulmar	Amerada Hess (1.10), Amoco (1.56), Texas Eastern (0.63), Enterprise Oil (1.56), Esso (46.97), Mobil (1.21), Shell (46.97)
15. Heather	DNO (6.25), Tenneco (31.25), Texaco (31.25), Union Oil (31.25)
16. Hutton	Amerada Hess (7.23), Amoco (10.31), Britoil (20.00), Conoco (20.00), Enterprise Oil (10.31), Gulf (20.00), Mobil (8.00), Texas Eastern (4.15)
17. NW Hutton	Amerada Hess (18.08), Amoco (25.77), Enterprise Oil (25.77), Mobil (20.00), Texas Eastern (10.38)

Table S.7: Continued

Fields, UK	*Licensees*
18. Magnus	BP (100.0)
19. Maureen	Agip (17.26), CPL (11.50), Petrofina (28.96), Phillips (33.78), Ultramar (8.50)
20. Montrose	Amerada Hess (23.08), Amoco (30.77), Enterprise Oil (30.77), Texas Eastern (15.38)
21. Murchison (UK)	Britoil (33.33), Conoco (33.33), Gulf (33.33)
22. Ninian	BP (14.38), Britoil (21.37), Chevron (17.10), ICI (18.52), Lasmo (8.63), Murphy/Odeco (7.12/7.12), Ranger (5.75)
23. Piper	Allied (20.0), Getty (23.5), ITO (20.0), Occidental (36.5)
24. Statfjord (UK)	Britoil (33.33), Conoco (33.33), Gulf (33.33)
25. Tartan	Texaco (100.0)
26. Thistle	Britoil (18.41), Burmah (8.40), Charterhouse (2.41), Deminex (41.50), Reading and Bates (1.00), Santa Fe (16.87), Tricentrol (10.00), Ultramar (1.41)

Fields, Norway	
1. Albuskjell	Agip (6.52), CFP (2.02), Cofranord (0.15), Coparex (0.2), Elf (4.05), Eurafrep (0.23), Norsk Hydro (3.35), Petrofina (15.0), Phillips (18.48), Shell (50.0)
2. Cod 3. Edda 4. Ekofisk 5. W Ekofisk 6. Eldfisk	Agip (13.04), CFP (4.05), Cofranord (0.3), Coparex (0.4), Elf (8.09), Eurafrep (0.46), Norsk Hydro (6.7), Petrofina (30.0), Phillips (36.96)
7. Tor	Agip (9.83), Amerada Hess (6.98), Amoco (6.98), CFP (3.05), Cofranord (0.22), Coparex (0.3), Elf (6.1), Eurafrep (0.35), Noco (3.7), Norsk Hydro (5.05), Petrofina (22.61), Phillips (27.85), Texas Eastern (6.98)
8. Murchison	Amerada Hess (1.042), Amoco (Norway) (1.042), Conoco (10.000), Esso (10.000), Mobil (15.000), Saga (1.875), Shell (10.000), Statoil (50.000), Texas Eastern (1.042)
9. Statfjord (Norway)	As Murchison
10. Valhall	Amerada Hess (28.1), Amoco (28.1), Noco (15.3), Statoil (0.4), Texas Eastern (28.1)

Table S.8:   UK and Norway. Companies' Licences in Producing Oilfields.
Percentage Shares. 1984.

Company		*Fields: (a) UK, (b) Norway*
1.	AB Exploration	(a) Claymore (0.50)
2.	Agip	(a) Claymore (2.50), Maureen (17.26)
		(b) Albuskjell (6.52), Ekofisk five fields (13.04), Tor (9.83)
3.	Allied	(a) Claymore (20.0), Piper (20.0)
4.	Amerada Hess	(a) Beryl (20.0), Fulmar (1.10), Hutton (7.23), NW Hutton (18.08), Montrose (23.08)
		(b) Murchison (1.042), Statfjord (1.042), Tor (6.98), Valhall (28.1)
5.	Amoco	(a) Fulmar (1.56), Hutton (10.31), NW Hutton (25.77), Montrose (30.77)
		(b) Murchison (1.042), Statfjord (1.042), Tor (6.98), Valhall (28.1)
6.	Anvil	(a) Claymore (0.5)
7.	Associated News	(a) Argyll (12.5), Duncan (12.5)
8.	BC Resources	(a) Brae (13.1)
9.	Berkeley Expl.	(a) Forties (0.25)
10.	Bow Valley	(a) Brae (4.2)
11.	BP	(a) Buchan (24.58), Forties (83.13), Magnus (100.0), Ninian (14.38)
12.	Britoil	(a) Beatrice (28.0), Brae (20.0), Dunlin (9.77), Hutton (20.0), Murchison (33.33), Ninian (21.37), Statfjord (33.33), Thistle (18.41)
13.	Burmah	(a) Thistle (8.40)
14.	CFP	(b) Albuskjell (2.02), Ekofisk five fields (4.05), Tor (3.05)
15.	Charterhall	(a) Buchan (4.14), Forties (0.25)
16.	Charterhouse	(a) Buchan (12.71), Forties (1.25), Thistle (2.41)
17.	Chevron	(a) Ninian (17.10)
18.	Clyde	(a) Buchan (12.71)
19.	Coalite	(a) Claymore (1.0)
20.	Cofranord	(b) Albuskjell (0.15), Ekofisk five fields (0.3), Tor (0.22)
21.	Conoco	(a) Dunlin (9.77), Hutton (20.0), Murchison (33.33), Statfjord (33.33)
		(b) Murchison (10.0), Statfjord (10.0)
22.	Coparex	(b) Albuskjell (0.2), Ekofisk five fields (0.4), Tor (0.3)
23.	CPL	(a) Forties (0.25), Maureen (11.5)
24.	Deminex	(a) Beatrice (22.0), Thistle (41.5)
25.	DNO	(a) Heather (6.25)
26.	Dow Chemical	(a) Claymore (5.0)
27.	Elf-Aquitaine	(a) Forties (1.5)
		(b) Albuskjell (4.05), Ekofisk five fields (8.09), Tor (6.1)

Table S.8: Continued

	Company		Fields: (a) UK, (b) Norway
28.	Enterprise Oil	(a)	Beryl (10.0), Fulmar (1.56), Hutton (10.31), NW Hutton (25.77), Montrose (30.77)
29.	Esso	(a)	Auk (50.0), Brent (50.0), Cormorant (50.0), Cormorant S (50.0), Dunlin (35.35), Forties (2.61), Fulmar (46.97)
		(b)	Murchison (10.0), Statfjord (10.0)
30.	Eurafrep	(b)	Albuskjell (0.23), Ekofisk five fields (0.46), Tor (0.35)
31.	Floyd	(a)	Claymore (0.5)
32.	Getty	(a)	Claymore (21.20),Piper (23.50)
33.	Goal Petroleum	(a)	Buchan (4.54)
34.	Gulf	(a)	Dunlin (9.77), Hutton (20.0), Murchison (33.33), Statfjord (33.33)
35.	Hamilton Oil	(a)	Argyll (36.0), Duncan (36.0)
36.	Hispanoil	(a)	Forties (0.25)
37.	Hunt	(a)	Beatrice (10.0)
38.	ICI	(a)	Ninian (18.52)
39.	Ind Scot Energy	(a)	Forties (0.25)
40.	ITO	(a)	Claymore (20.0), Piper (20.0)
41.	Kerr McGee	(a)	Beatrice (25.0), Brae (8.0)
42.	Kleinwort Benson	(a)	Argyll (2.5), Duncan (2.5)
43.	Lasmo	(a)	Beatrice (15.0), Ninian (8.63)
44.	Lochiel	(a)	Buchan (0.9)
45.	Louisiana Land	(a)	Brae (10.7)
46.	Marathon	(a)	Brae (38.0)
47.	Mobil	(a)	Beryl (50.0), Fulmar (1.21), Hutton (8.0), NW Hutton (20.0)
		(b)	Murchison (15.0), Statfjord (15.0)
48.	Moray Petroleum	(a)	Beatrice (2.50)*
49.	Murphy	(a)	Ninian (7.12)
50.	Noco	(b)	Tor (3.7), Valhall (15.3)
51.	Norsk Hydro	(a)	Brae (2.0), Forties (0.25)
		(b)	Albuskjell (3.35), Ekofisk five fields (6.7), Tor (5.05)
52.	North Sea & General	(a)	Claymore (1.0), Forties (0.25)
53.	Norwich Union	(a)	Forties (0.25)
54.	Occidental	(a)	Claymore (23.40), Piper (36.50)
55.	Odeco	(a)	Ninian (7.12)
56.	OK Exploration	(a)	Forties (0.95)
57.	Petrofina	(a)	Maureen (28.96)
		(b)	Albuskjell (15.0), Ekofisk five fields (30.0), Tor (22.61)
58.	Phillips	(a)	Maureen (33.78)
		(b)	Albuskjell (18.48), Ekofisk five fields (36.96), Tor (27.85)
59.	Pict	(a)	Claymore (0.5)

Table S.8: Continued

	Company		*Fields: (a) UK, (b) Norway*
60.	Plascom	(a)	Forties (0.25)
61.	Ranger	(a)	Ninian (5.75)
62.	Reading & Bates	(a)	Thistle (1.0)
63.	RTZ	(a)	Argyll (25.0), Duncan (25.0), Forties (1.0)
64.	Saga	(b)	Murchison (1.875), Statfjord (1.875)
65.	Santa Fe	(a)	Thistle (16.87)
66.	Saxon	(a)	Forties (0.25)
67.	Shell	(a)	Auk (50.0), Brent (50.0), Cormorant (50.0), Cormorant S (50.0), Dunlin (35.35), Forties (2.61), Fulmar (46.97)
		(b)	Albuskjell (50.0), Murchison (10.0), Statfjord (10.0)
68.	Sovereign	(a)	Brae (4.0), Claymore (2.0), Forties (0.5)
69.	Statoil	(b)	Murchison (50.0), Statfjord (50.0), Valhall (0.4)*
70.	Sulpetro	(a)	Buchan (12.71)
71.	Tenneco	(a)	Heather (31.25)
72.	Texaco	(a)	Argyll (24.0), Buchan (9.23), Duncan (24.0), Forties (1.0), Heather (31.25),Tartan (100.0)
73.	Texas Eastern	(a)	Beryl (20.0), Fulmar (0.63), Hutton (4.15), NW Hutton (10.38), Montrose (15.38)
		(b)	Murchison (1.042), Statfjord (1.042), Tor (7.0), Valhall (28.1)
74.	Texas Gas	(a)	Claymore (0.6)
75.	Third Triton	(a)	Claymore (0.5)
76.	Trafalgar House	(a)	Forties (1.45)
77.	Transworld	(a)	Buchan (12.71), Claymore (0.8)
78.	Tricentrol	(a)	Buchan (5.76), Thistle (10.0)
79.	Ultramar	(a)	Forties (1.0), Maureen (8.5), Thistle (1.41)
80.	Union Jack	(a)	Forties (0.25)
81.	Union Oil	(a)	Heather (31.25)
82.	Viva	(a)	Forties (0.25)

* net profit interest

Table S.9:   UK and Norway. Operators and Licensing Rounds of Producing
Oilfields. 1984.

Field	Operator	Licensing Round
Argyll	Hamilton	2
Auk	Shell	3
Beatrice	Britoil	4
Beryl	Mobil	4
S Brae	Marathon	3
Brent	Shell	3
Buchan	BP	4
Claymore	Occidental	4
Cormorant	Shell	4
Cormorant S	Shell	4
Deveron	Britoil	4
Duncan	Hamilton	2
Dunlin	Shell	4
Ekofisk Area (Norway)	Phillips	2
Forties	BP	2
Fulmar	Shell	3
Heather	Union	4
Hutton	Conoco	4
NW Hutton	Amoco	4
Magnus	BP	4
Maureen	Phillips	3
Montrose	Amoco	1
Murchison (UK and Norway)	Conoco	3
Ninian	Chevron	4
Piper	Occidental	4
Statfjord (Norway)	Mobil	3
Statfjord (UK)	Conoco	3
Tartan	Texaco	4
Thistle	Britoil	4
Valhall (Norway)	Amoco	2

Operator	Field
1. Amoco	NW Hutton, Montrose, Valhall (Norway)
2. BP	Buchan, Forties, Magnus
3. Britoil	Beatrice, Deveron, Thistle
4. Chevron	Ninian
5. Conoco	Hutton, Murchison (UK and Norway), Statfjord (UK)
6. Hamilton	Argyll, Duncan
7. Marathon	S Brae
8. Mobil	Beryl, Statfjord (Norway)
9. Occidental	Claymore, Piper
10. Phillips	Ekofisk Area (Norway), Maureen
11. Shell	Auk, Brent, Cormorant, Cormorant S, Dunlin, Fulmar
12. Texaco	Tartan
13. Union	Heather

Table S.10: UK and Norway. Term (BNOC/Statoil) Prices of North Sea Crudes. 1980–85. Dollars.

	Brent	*Forties*	*Flotta*	*Ninian*	*Ekofisk*
1980					
Jan 1	29.75	29.75	29.00	29.55	32.50
Feb 4	33.75	33.75	33.00	33.55	34.50
Mar					
Apr 1	34.25	34.25	33.00	33.95	35.00
May 20	36.25	36.25	35.00	35.95	37.00
Jun					
Jul					
Aug					
Sep					
Oct					
Nov					
Dec					
1981					
Jan 1	39.25	39.25	38.25	38.95	40.00
Feb					
Mar					
Apr					
May					
Jun 15	35.00	35.00	34.00	34.70	
Jun 20					35.75
Jul					
Aug					
Sep					
Oct					
Nov 1	36.60	36.50	35.50	36.10	37.25
Dec					
1982					
Jan 1			35.25		
Feb 9	35.10	35.00	33.75	34.60	35.75
Mar 1	31.10	31.00	30.00	30.60	
Apr					
May					
Jun 1	33.50	33.50	32.50	33.10	34.25
Jul					
Aug					
Sep					
Oct					
Nov					
Dec					

Table S.10: Continued

	Brent	Forties	Flotta	Ninian	Ekofisk
1983					
Jan					
Feb 1	30.50	30.50	29.55	30.10	31.00
Mar 1	30.00	29.75	28.80	29.35	30.25
Apr					
May					
Jun					
Jul					
Aug					
Sep					
Oct 1		29.90	29.30	29.60	
Nov					
Dec					
1984					
Jan					
Feb					
Mar					
Apr 1					30.10
May					
Jun					
Jul					
Aug					
Sep					
Oct 1					28.65
Oct 17	28.65	28.55	27.95	28.40	
Nov 1					28.95
Dec					
1985					
Jan					
Feb					
Mar					
Apr 1	27.50	27.50	27.00	27.35	
May 1	27.90	27.90	27.40	27.75	
Jun 1	26.65	26.65	26.15	26.50	

Source: *PIW*

Table S.11: Monthly Spot Prices of Selected Crudes. 1980–85. Dollars.

	Arabian Light	Arabian Heavy	Forcados	Urals	WTI
1980					
Jan	37.58	34.47	39.50	38.25	
Feb	35.33	32.40	n.a.	36.05	
Mar	36.25	32.00	n.a.	36.38	
Apr	35.33	32.75	37.00	36.00	
May	35.96	33.75	37.50	36.00	
Jun	35.78	n.a.	37.40	35.50	
Jul	34.55	32.81	36.13	35.00	
Aug	32.10	30.55	32.25	32.63	
Sep	32.13	30.56	32.50	33.50	
Oct	37.50	35.45	37.50	37.20	
Nov	41.25	39.25	41.31	40.81	
Dec	39.31	37.63	38.75	40.05	
1981					
Jan	38.95	37.25	39.75	39.44	
Feb	37.00	35.25	37.88	37.75	
Mar	36.69	34.94	37.75	37.31	
Apr	36.44	34.00	37.06	36.83	
May	33.55	31.35	34.95	34.45	
Jun	31.81	29.75	33.75	32.81	
Jul	31.84	29.40	34.35	32.60	
Aug	32.21	29.31	34.88	32.83	
Sep	31.98	28.94	34.74	32.40	
Oct	33.63	29.13	35.82	33.75	
Nov	34.28	30.69	36.81	34.75	
Dec	34.14	30.28	36.38	34.38	
1982					
Jan	34.04	29.87	35.76	34.38	
Feb	30.16	26.53	31.38	30.53	
Mar	28.44	25.81	28.94	29.20	
Apr	30.65	27.90	31.90	32.25	
May	33.44	30.30	34.69	33.38	
Jun	32.88	30.60	34.44	32.72	
Jul	31.70	29.80	33.70	32.38	
Aug	31.26	29.33	33.26	32.00	
Sep	33.10	30.73	34.19	33.31	
Oct	33.45	30.80	34.91	32.81	
Nov	31.69	29.55	34.19	32.25	33.26
Dec	30.26	28.40	32.15	31.40	31.28

Table S.11: Continued

	Arabian Light	Arabian Heavy	Forcados	Urals	WTI
1983					
Jan	30.44	28.58	31.19	31.25	30.58
Feb	29.15	27.44	28.88	28.69	28.24
Mar	28.00	26.20	27.50	27.88	28.79
Apr	28.60	26.15	28.67	28.75	30.56
May	28.56	25.98	29.20	28.88	30.10
Jun	28.80	26.25	29.53	29.28	30.99
Jul	28.94	26.50	30.21	29.71	31.28
Aug	28.96	26.70	30.65	29.93	31.78
Sep	28.65	26.63	30.58	29.65	31.24
Oct	28.61	26.59	29.66	29.44	30.34
Nov	28.23	26.28	29.44	28.91	29.98
Dec	28.26	26.22	28.98	28.73	29.26
1984					
Jan	28.64	26.35	29.40	28.78	29.75
Feb	28.51	26.56	29.63	29.24	30.05
Mar	28.53	26.80	29.85	29.42	30.81
Apr	28.39	26.89	29.89	29.38	30.53
May	28.38	27.04	29.75	29.20	30.59
Jun	28.07	26.94	29.57	29.97	29.73
Jul	27.43	26.56	28.21	28.48	28.83
Aug	27.71	26.69	28.33	27.71	29.25
Sep	27.81	27.00	28.51	28.00	29.38
Oct	27.70	26.61	28.44	28.20	28.43
Nov	27.92	26.58	27.89	28.08	28.05
Dec	27.68	26.16	27.33	27.76	26.13
1985					
Jan	27.94	26.61	27.16	27.94	25.53
Feb	27.80	26.68	28.21	28.56	
Mar	27.81	26.74	28.25	28.42	
Apr	27.65	26.40	28.13	27.78	
May	26.80	25.56	26.97	26.55	
Jun	26.83	25.13	26.50	25.55	
Jul	27.05	25.11	26.54	25.96	
Aug	27.40	25.29	27.55	26.61	
Sep	27.61	25.55	28.25	27.41	
Oct	27.86	25.63	28.85	27.80	
Nov	27.89	25.76	29.79		
Dec	27.75	25.71	27.36		

Sources: *OPEC Bulletin* (Urals)
International Crude Oil and Product Prices, Middle East Petroleum and Economic Publications (Arabian Light, Arabian Heavy, Forcados)
Industry source (WTI)

Table S.12: Monthly Spot Prices of North Sea Crudes. 1980–85. Dollars.

	Brent	*Flotta*	*Ekofisk*
1980			
Jan	39.43	39.13	38.50
Feb	37.44	36.40	39.10
Mar	37.19	36.63	38.50
Apr	36.94	36.75	38.50
May	36.94	36.95	38.63
Jun	37.19	36.44	38.25
Jul	35.44	35.45	36.44
Aug	31.95	32.00	33.55
Sep	31.70	33.00	33.38
Oct	36.44	37.50	38.30
Nov	41.93	40.94	42.25
Dec	40.93	40.30	39.75
1981			
Jan	39.18	39.50	40.45
Feb	37.93	37.44	38.75
Mar	37.64	37.19	38.25
Apr	37.64	36.53	37.13
May	33.44	33.60	34.85
Jun	32.19	32.25	33.13
Jul	35.19	32.65	34.80
Aug	34.94	32.67	35.94
Sep	34.84	32.40	35.65
Oct	35.44	34.13	36.47
Nov	37.04	34.81	37.13
Dec	36.54	34.38	36.75
1982			
Jan	35.94	34.31	35.95
Feb	29.55	30.21	31.56
Mar	27.95	28.90	28.88
Apr	31.95	32.31	32.75
May	34.94	33.50	35.25
Jun	34.01	32.97	34.90
Jul	33.19	32.25	33.77
Aug	32.33	31.69	32.65
Sep	34.08	33.56	34.25
Oct	34.43	33.31	35.11
Nov	33.03	32.00	33.56
Dec	31.13	31.05	31.70

Table S.12: Continued

	Brent	Flotta	Ekofisk
1983			
Jan	30.72	30.56	31.06
Feb	28.62	28.44	29.13
Mar	28.03	27.56	28.48
Apr	29.70	28.50	29.71
May	29.47	28.63	29.70
Jun	30.23	29.18	30.21
Jul	30.79	29.50	30.81
Aug	31.14	29.44	31.31
Sep	30.37	28.90	30.64
Oct	29.70	28.75	29.80
Nov	29.02	28.60	29.11
Dec	28.87	28.30	28.86
1984			
Jan	29.48	28.80	29.56
Feb	29.70	29.20	29.88
Mar	30.05	29.63	30.08
Apr	29.95	29.63	30.13
May	29.80	29.41	29.90
Jun	28.78	28.87	29.20
Jul	27.78	28.17	27.91
Aug	28.05	27.80	28.13
Sep	28.31	27.93	28.38
Oct	28.06	27.54	28.01
Nov	27.62	27.52	27.74
Dec	26.98	26.79	26.96
1985			
Jan	27.04	26.75	27.01
Feb	28.36	28.34	28.50
Mar	28.07	27.88	27.93
Apr	27.91	27.81	27.85
May	26.74	26.54	26.73
Jun	26.60	26.20	26.53
Jul	26.73	26.46	26.83
Aug	27.41	26.94	27.49
Sep	27.80	27.55	27.89
Oct	28.74	28.28	28.73
Nov	29.70		29.78
Dec	26.79		26.88

Sources: *OPEC Bulletin* (Flotta)
International Crude Oil and Product Prices, Middle East Petroleum and
Economic Publications (Ekofisk; Brent, March–December 1985)
Industry source (Brent, January 1980–June 1982)
Petroleum Argus (Brent, July 1982–February 1985)

Table S.13: Brent Market Prices: Weekly Averages. Dollars.

Week		Current Price	One-month Price	Two-month Price	Three-month Price	Four-month Price	Five-month Price
1983							
Week	31	–	31.26	–			
	32	31.23	31.30	31.30			
	33	31.01	30.90	–			
	34	31.02	31.05	–			
	35	30.95	30.85	–			
	36	30.25	30.47	–			
	37	30.28	30.43	30.30			
	38	30.53	30.54	–			
	39	29.94	29.95	–			
	40	29.21	29.55	–			
	41	29.83	29.98	–			
	42	29.83	29.70	–			
	43	29.82	29.72	–			
	44	29.45	29.58	–			
	45	29.17	29.29	29.02			
	46	28.62	28.71	–			
	47	28.64	28.83	–			
	48	28.78	28.88	28.76			
	49	28.70	28.61	28.52			
	50	28.80	28.70	28.53			
	51	28.74	28.68	28.14			
	52	29.58	29.29	29.33			
1984							
Week	1	29.41	29.16	28.85			
	2	29.26	29.15	29.01			
	3	29.48	29.32	29.13			
	4	29.76	29.51	29.35			
	5	30.02	29.73	29.45			
	6	29.84	29.48	29.18			
	7	29.46	29.34	29.01			
	8	29.68	29.32	29.09	–		
	9	29.89	29.89	29.66	29.50		
	10	30.10	30.06	29.96	29.68		
	11	30.25	30.02	29.99	29.73		
	12	29.94	29.69	29.50	–		
	13	30.14	29.93	29.86	29.75		
	14	30.25	29.96	29.88	29.74		
	15	30.10	30.05	29.95	29.89	29.83	
	16	29.95	29.78	29.70	–	–	
	17	–	29.80	29.68	–	29.60	

Table S.13: Continued

Week		Current Price	One-month Price	Two-month Price	Three-month Price	Four-month Price	Five-month Price
	18	29.59	29.45	29.38	–	–	
	19	29.51	29.57	29.65	29.71	–	
	20	30.25	30.45	30.33	–	–	
	21	29.82	29.87	29.83	29.83	–	
	22	29.90	30.00	30.10	30.12	–	
	23	29.49	29.80	29.82	29.80	–	
	24	29.25	29.36	29.47	29.50	–	
	25	28.90	29.05	29.04	28.92	–	
	26	28.21	28.75	29.00	29.21	–	
	27	28.44	29.00	29.26	29.35	–	
	28	28.40	28.70	28.89	29.11	29.15	
	29	28.13	28.29	28.57	28.48	–	
	30	26.87	27.46	27.76	27.83	–	
	31	26.64	27.37	27.91	28.29	28.91	
	32	28.01	28.41	28.87	29.02	29.21	
	33	28.05	28.39	28.75	28.96	–	
	34	28.48	28.93	29.11	29.17	–	
	35	28.15	28.85	29.09	–	–	
	36	27.93	28.36	28.68	28.86	–	
	37	28.20	28.48	28.92	29.11	–	
	38	28.49	28.97	29.07	–	–	
	39	28.51	29.02	29.02	29.16	–	
	40	28.56	28.90	29.03	29.09	–	
	41	28.38	28.57	28.73	28.74	–	
	42	27.25	27.52	27.86	–	–	
	43	27.77	28.01	28.09	27.88	–	
	44	27.82	28.01	28.00	27.45	–	
	45	27.89	28.19	28.19	28.00	–	
	46	27.87	28.01	27.87	27.79	–	
	47	27.65	27.43	27.24	27.27	–	
	48	27.17	27.12	27.01	26.34	–	
	49	27.34	27.39	27.33	27.24	–	
	50	27.31	27.13	27.04	26.84	–	
	51	26.85	26.57	26.43	–	–	
	52	26.86	26.80	26.49	–	–	
1985							
Week	1	26.40	25.99	25.97	25.88	–	
	2	26.91	26.70	26.51	26.29	–	
	3	27.12	26.43	25.91	–	–	
	4	26.90	26.03	25.62	25.15	–	
	5	27.69	26.58	25.88	26.40	–	
	6	28.45	27.47	26.64	26.40	–	

Table S.13: Continued

Week		Current Price	One-month Price	Two-month Price	Three-month Price	Four-month Price	Five-month Price
	7	29.00	28.25	27.24	26.74	–	
	8	28.71	27.35	26.87	26.55	–	
	9	27.41	26.47	26.18	25.82	–	
	10	28.52	27.31	26.77	26.59	–	
	11	28.60	27.92	26.90	26.72	–	
	12	28.11	27.31	27.04	–	–	
	13	28.25	27.37	27.04	–	–	
	14	28.38	27.72	27.36	27.27	–	
	15	28.72	28.07	27.64	27.40	27.19	
	16	28.01	27.43	27.16	27.00	–	
	17	27.61	27.16	26.86	26.78	26.67	
	18	27.02	26.55	26.48	26.40	26.44	
	19	26.51	26.20	26.24	26.30	26.29	
	20	27.10	26.41	26.30	26.25	26.23	
	21	26.64	26.35	26.35	26.35	26.31	
	22	27.02	26.72	26.62	26.60	–	
	23	26.82	26.26	26.08	26.01	26.02	
	24	26.62	26.20	25.52	25.40	25.44	
	25	26.23	25.64	25.43	25.21	25.35	
	26	26.57	26.12	25.84	25.66	25.69	25.25

Week		Current Price	One-month Price	Two-month Price	Three-months plus Price
1985					
Week	27	26.73	26.12	25.88	25.31
	28	26.93	26.39	26.01	25.73
	29	26.55	25.85	25.53	25.32
	30	26.98	26.31	25.85	25.48
	31	26.61	26.52	25.98	25.86
	32	26.67	26.57	26.13	25.83
	33	26.64	26.94	27.71	26.30
	34	27.47	27.15	26.96	26.66
	35	27.72	27.46	27.15	27.07
	36	27.53	27.41	27.04	26.85
	37	26.92	26.95	26.47	26.22

Note: For weeks 27–37 of 1985, four- and five-month deals are included in the calculations of average three-months plus prices.

Source: *Petroleum Argus*

Table S.14: Brent Market Activity: Total Number of Deals per Week

Week		Current Month Deals	One-month Deals	Two-month Deals	Three-month Deals	Four-month Deals	Five-month Deals
1983							
Week	31	–	10	–			
	32	4	6	1			
	33	4	1	–			
	34	2	1	–			
	35	3	1	–			
	36	2	9	–			
	37	2	8	1			
	38	22	4	–			
	39	11	14	–			
	40	5	20	–			
	41	3	22	–			
	42	6	2	–			
	43	9	10	–			
	44	12	12	–			
	45	10	17	3			
	46	10	4	–			
	47	8	3	–			
	48	6	12	4			
	49	4	14	4			
	50	1	7	6			
	51	14	14	4			
	52	6	14	2			
1984							
Week	1	7	6	8			
	2	4	10	14			
	3	15	7	3			
	4	25	23	1			
	5	9	10	2			
	6	3	21	5			
	7	2	7	9			
	8	25	31	6			
	9	9	29	12	5		
	10	9	22	11	2		
	11	1	5	6	4		
	12	11	14	8	–		
	13	14	15	7	2		
	14	2	15	5	3		
	15	1	9	18	9	1	
	16	5	12	2	–	–	
	17	–	24	4	–	2	

Table S.14: Continued

Week		Current Month Deals	One-month Deals	Two-month Deals	Three-month Deals	Four-month Deals	Five-month Deals
	18	13	39	6	–	–	
	19	4	20	31	3	–	
	20	9	22	8	–	–	
	21	10	34	3	1	–	
	22	3	8	8	3	–	
	23	3	8	12	2	–	
	24	5	15	16	14	–	
	25	21	26	31	9	–	
	26	11	23	11	9	–	
	27	4	12	21	14	–	
	28	9	22	20	18	1	
	29	12	24	40	12	–	
	30	6	28	33	27	–	
	31	13	31	23	19	3	
	32	8	20	20	15	3	
	33	2	9	10	9	–	
	34	16	31	24	8	–	
	35	19	15	21	–	–	
	36	11	18	27	21	–	
	37	7	11	18	14	–	
	38	22	22	11	–	–	
	39	14	20	10	2	–	
	40	11	12	15	5	–	
	41	7	30	28	10	–	
	42	14	76	10	–	–	
	43	14	73	13	2	–	
	44	7	51	7	2	–	
	45	9	24	9	2	–	
	46	3	12	3	5	–	
	47	34	24	11	1	–	
	48	16	19	20	9	–	
	49	9	15	21	11	–	
	50	4	17	25	14	–	
	51	28	24	8	–	–	
	52	2	2	6	–	–	

1985

Week	1	9	20	15	4	–	
	2	6	52	37	4	–	
	3	29	16	2	–	–	
	4	37	50	3	2	–	
	5	21	79	28	1	–	
	6	8	58	29	1	–	

Table S.14: Continued

Week	Current Month Deals	One-month Deals	Two-month Deals	Three-month Deals	Four-month Deals	Five-month Deals
7	1	60	44	11	–	
8	28	25	26	2	–	
9	20	49	30	2	–	
10	16	54	34	7	–	
11	2	33	27	11	–	
12	20	37	31	–	–	
13	12	25	10	–	–	
14	11	32	21	7	–	
15	7	30	35	14	4	
16	19	16	14	1	–	
17	21	58	8	10	7	
18	15	45	19	12	4	
19	6	54	21	8	9	
20	1	9	12	1	3	
21	30	32	14	2	2	
22	19	31	12	10	–	
23	10	46	44	16	14	
24	1	31	35	20	5	
25	27	24	12	8	1	
26	10	26	14	6	1	1

Week	Current Month Deals	One-month Deals	Two-month Deals	Three-months plus Deals

1985

Week				
27	9	10	8	6
28	10	43	18	7
29	26	17	16	24
30	18	32	13	18
31	12	32	24	22
32	14	21	19	23
33	11	32	33	35
34	28	26	18	14
35	25	29	19	12
36	9	29	25	18
37	16	57	30	28

Note: For weeks 27–37 of 1985, four- and five-month deals are included in the number of three-months plus deals recorded.

Source: *Petroleum Argus*

Table S.15: UK. Payback Period for Producing Fields. (Post-tax Basis Calculation of Number of Years of Income Required to Recoup Field Investment.)

Field	Payback Period	Field	Payback Period
Argyll	2.0	Heather	6.0
Auk	2.0	Hutton	3.0
Beatrice	7.0	NW Hutton	2.0
Beryl	4.0	Magnus	3.0
Brae	8.0	Maureen	4.0
Brent	7.0	Montrose	4.0
Buchan	2.0	Murchison (UK)	3.0
Claymore	4.0	Ninian	4.0
Cormorant Area	7.0	Piper	2.0
Duncan	3.0	Statfjord (UK)	5.0
Dunlin	2.0	Tartan	12.0
Forties	3.0	Thistle	3.0
Fulmar	1.0		